Cambridge Lower Secondary Complete Chemistry

Philippa Gardom Hulme

Anna Harris

Onn May Ling

Second Edition

Contents

Introduction		4
Thinking and working scientifically		
1	Asking questions	8
2	Planning and carrying out investigations	10
3	Collecting and recording data	12
4	Drawing graphs	14
5	Analysis	16
6	Evaluation	18

Stage 7

1 The particle model
1.1	The particle model	20
1.2	The states of matter	22
1.3	Using the particle model	24
1.4	Changes of state – evaporating, boiling, and condensing	26
1.5	Investigating boiling points	28
1.6	Changes of state – melting and freezing	30
1.7	Models in science	32
1.8	**Review**	34

2 Elements, compounds, and mixtures
2.1	Elements and the periodic table	36
2.2	Discovering the elements	38
2.3	Chemical symbols	40
2.4	Atoms	42
2.5	Organising the elements	44
2.6	Compounds	46
2.7	What's in a name?	48
2.8	Chemical formulae	50
2.9	What's in a mixture?	52
2.10	What's in a solution?	54
2.11	Comparing elements, mixtures, and compounds	56
2.12	What are you made of?	58
2.13	**Review**	60

3 Metals and non-metals
3.1	Magnificent metals	62
3.2	Comparing conductors	64
3.3	Amazing alloys	66
3.4	Non-metal elements	68
3.5	Explaining metal and non-metal properties	70
3.6	Better bicycles	72
3.7	**Review**	74

4 Chemical reactions 1
4.1	What are chemical reactions?	76
4.2	Atoms in chemical reactions	78
4.3	Investigating a chemical reaction	80
4.4	Precipitation reactions	82
4.5	Corrosion reactions	84
4.6	**Review**	86

5 Acids and alkalis
5.1	Acids and alkalis	88
5.2	The pH scale	90
5.3	Neutralisation reactions	92
5.4	investigating neutralisation	94
5.5	Acid rain	96
5.6	Gas products of acid reactions	98
5.7	**Review**	100

6 Models of the Earth
6.1	Models of the Earth	102
6.2	Plate tectonics	104
6.3	The restless Earth	106
6.4	Volcanoes	108
6.5	**Review**	110

Stage 7 review 112

Stage 8

7 Inside atoms
7.1	Inside atoms	114
7.2	Discovering electrons	116
7.3	Finding the nucleus	118
7.4	Inside sub-atomic particles	120
7.5	Proton and nucleon numbers	122
7.6	**Review**	124

Contents

8	**Pure substances and solutions**	
8.1	Pure substances	126
8.2	Drinking seawater	128
8.3	Chromatography	130
8.4	Solutions and concentration	132
8.5	How much salt is in the sea?	134
8.6	Chlorine and water	136
8.7	Solubility	138
8.8	Investigating solubility and temperature – doing an investigation	140
8.9	Investigating temperature and solubility – writing up an investigation	142
8.10	**Review**	144

9	**Chemical reactions 2**	
9.1	More chemical reactions	146
9.2	Word equations	148
9.3	Energy changes	150
9.4	Investigating fuels	152
9.5	Investigating food energy	154
9.6	Metals and oxygen	156
9.7	Metals and water	158
9.8	Metals and acids	160
9.9	The reactivity series	162
9.10	Lead in the reactivity series	164
9.11	**Review**	166
Stage 8 review		168

Stage 9

10	**Structure, bonding, and properties**	
10.1	Proton number and the periodic table	170
10.2	Electrons in atoms	172
10.3	Making ions	174
10.4	Inside ionic compounds	176
10.5	Covalent bonding	178
10.6	Covalent structures	180
10.7	More about structures	182
10.8	Life-saving compounds	184
10.9	**Review**	186

11	**Patterns in the periodic table**	
11.1	Calculating density	188
11.2	Explaining density	190
11.3	Using density	192
11.4	The periodic table: Group 1	194
11.5	More about Group 1	196
11.6	The periodic table: Group 2	198
11.7	**Review**	200

12	**Chemical reactions 3**	
12.1	Mass and energy in chemical reactions	202
12.2	Writing symbol equations	204
12.3	Metal displacement reactions	206
12.4	Extracting metals	208
12.5	Extracting copper	210
12.6	Making salts from acids and metals	212
12.7	More about salts	214
12.8	Making salts from acids and carbonates	216
12.9	Rates of reaction	218
12.10	Concentration and reaction rate	220
12.11	Temperature and reaction rate	222
12.12	Surface area and reaction rate	224
12.13	**Review**	226

13	**Planet Earth**	
13.1	Continental drift	228
13.2	Evidence from fossils	230
13.3	Evidence from seafloor spreading	232
13.4	**Review**	234
Stage 9 review		236

Reference

1	Choosing apparatus	238
2	Working accurately and safely	240

Glossary 242
Index 246
The periodic table of the elements 256

Introduction

How to use your Student Book

Welcome to your Cambridge Lower Secondary Complete Chemistry Student Book. This book has been written to help you study Chemistry at all three stages of the Cambridge Lower Secondary Science curriculum framework.

Most of the units in this book work like this:

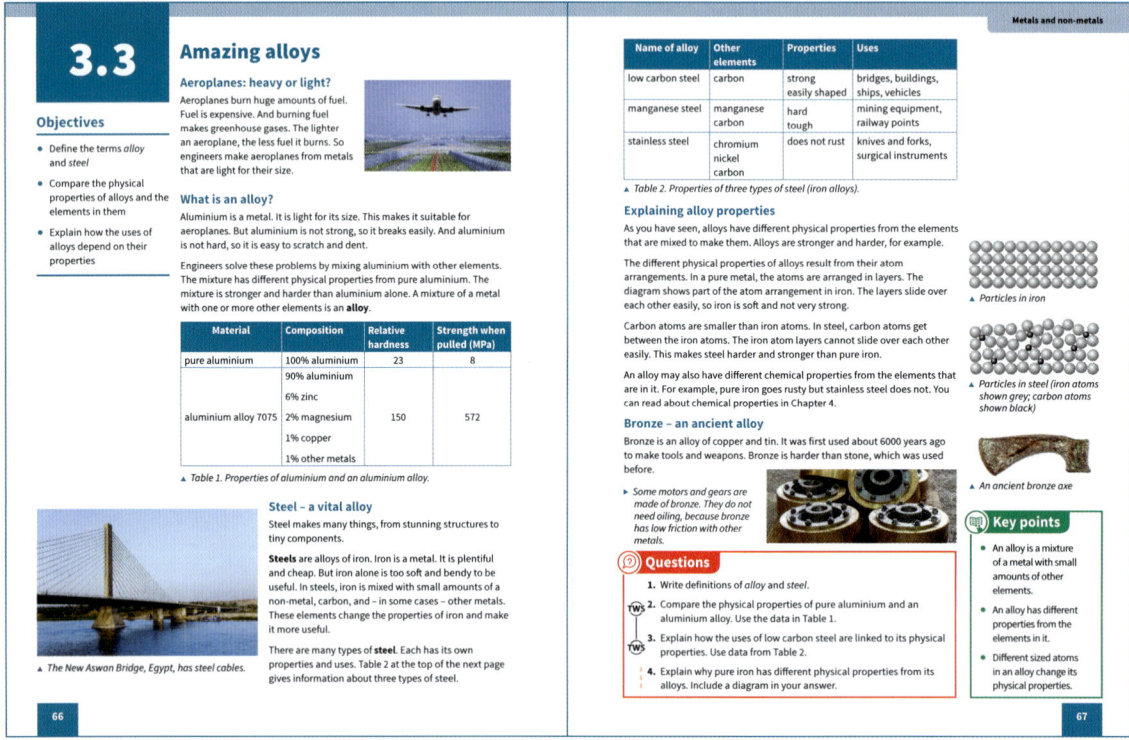

- Every page starts with the learning objectives for the unit. The learning objectives are linked to the Cambridge Lower Secondary Science curriculum framework.

- Key words are marked in **bold**. You can check the meaning of these words in the glossary at the back of the book.

- At the end of each unit there are questions to test that you understand what you have learned. The first question is straightforward and later questions are more challenging. The questions are written in the style of the Cambridge Checkpoint test, to help you prepare. Answers are available in the Teacher Handbook which is available in print and digitally via Kerboodle.

- The key points to remember from the unit are also summarised here.

These units cover the Chemistry topics in the Cambridge Lower Secondary Science curriculum framework.

In addition, many of the units help you think and work scientifically, put science in context, prepare for the next level, and test your knowledge.

Introduction

Thinking and working scientifically

Thinking and working scientifically is an important component of the curriculum framework.

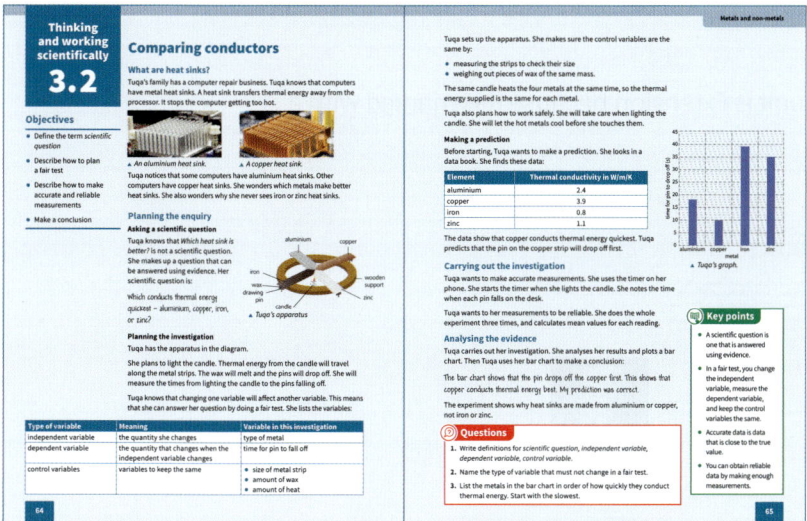

The Thinking and working scientifically units and features will help you learn:

- how to understand and apply models and representations
- the importance of asking scientific questions and planning how to answer them
- how to carry out enquiries such as fair test investigations and field work
- how to analyse data, draw conclusions, and evaluate your enquiry.

TWS Questions which test your Thinking and working scientifically skills and knowledge are marked with this icon.

On pages 8-19 you will find a dedicated Thinking and working scientifically chapter which introduces essential skills which will be useful throughout every stage of the curriculum framework.

Science in context

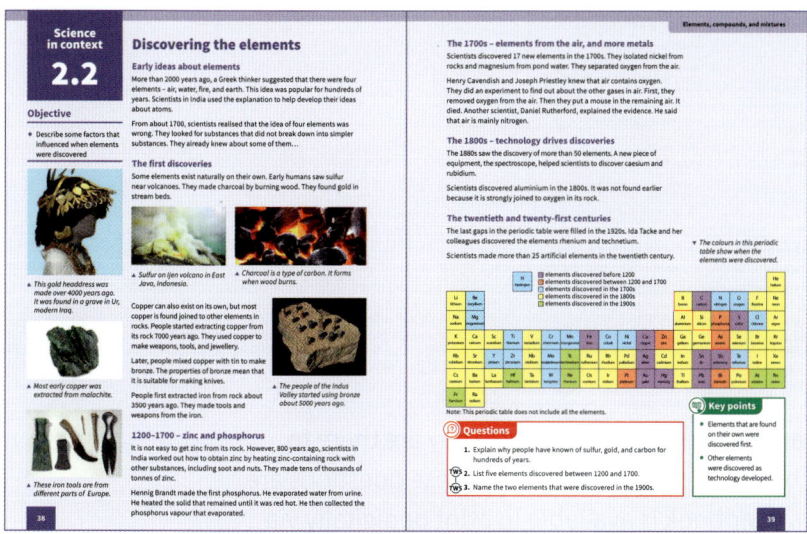

Science in context units will also help you learn:

- how scientists throughout history and from around the globe developed theories, carried out research, and drew conclusions about the world around them
- how science is applied in everyday life
- how issues involving chemistry are evaluated
- the global impact of the use of chemistry.

5

Extension

Throughout this book there are lots of opportunities to learn even more about chemistry, beyond the curriculum framework. These units are called Extension because they extend and develop your science skills further.

> You can tell when a question or part of a unit is Extension because it is marked with a dashed line, like the one on the left.

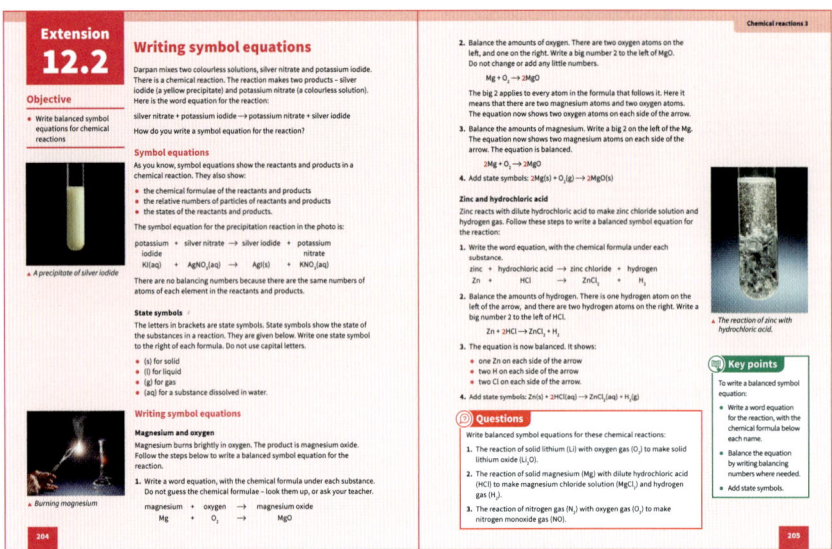

Extension units will not be part of your assessment, but they will help you prepare for moving onto the next stage of the curriculum and eventually for Cambridge IGCSE Chemistry.

Review

At the end of every chapter and every stage there are review questions.

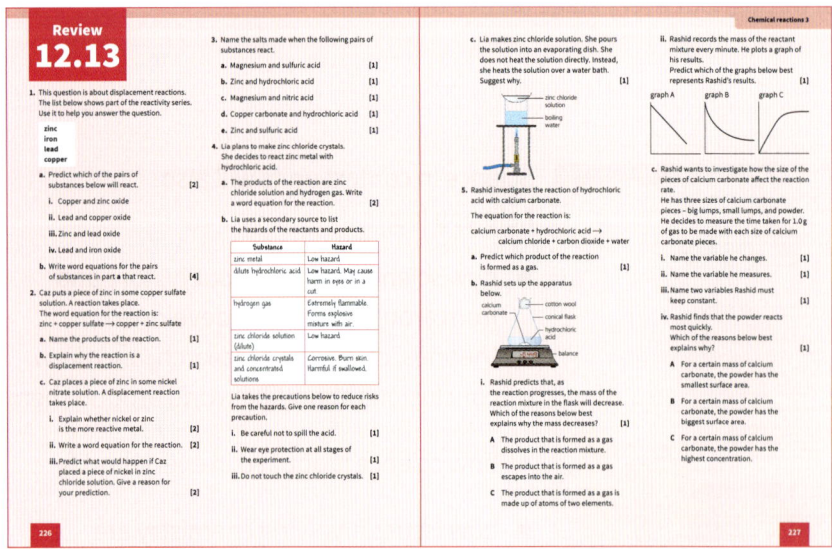

These questions are written in the style of the Cambridge Checkpoint test. They are there to help you review what you have learned in that chapter or stage. Answers to these questions are available in the Teacher Handbook. The Teacher Handbook is available in print or digitally via Kerboodle.

Reference

At the back of this book, on pages 238-241, there are reference pages providing further information that will help you while you study.

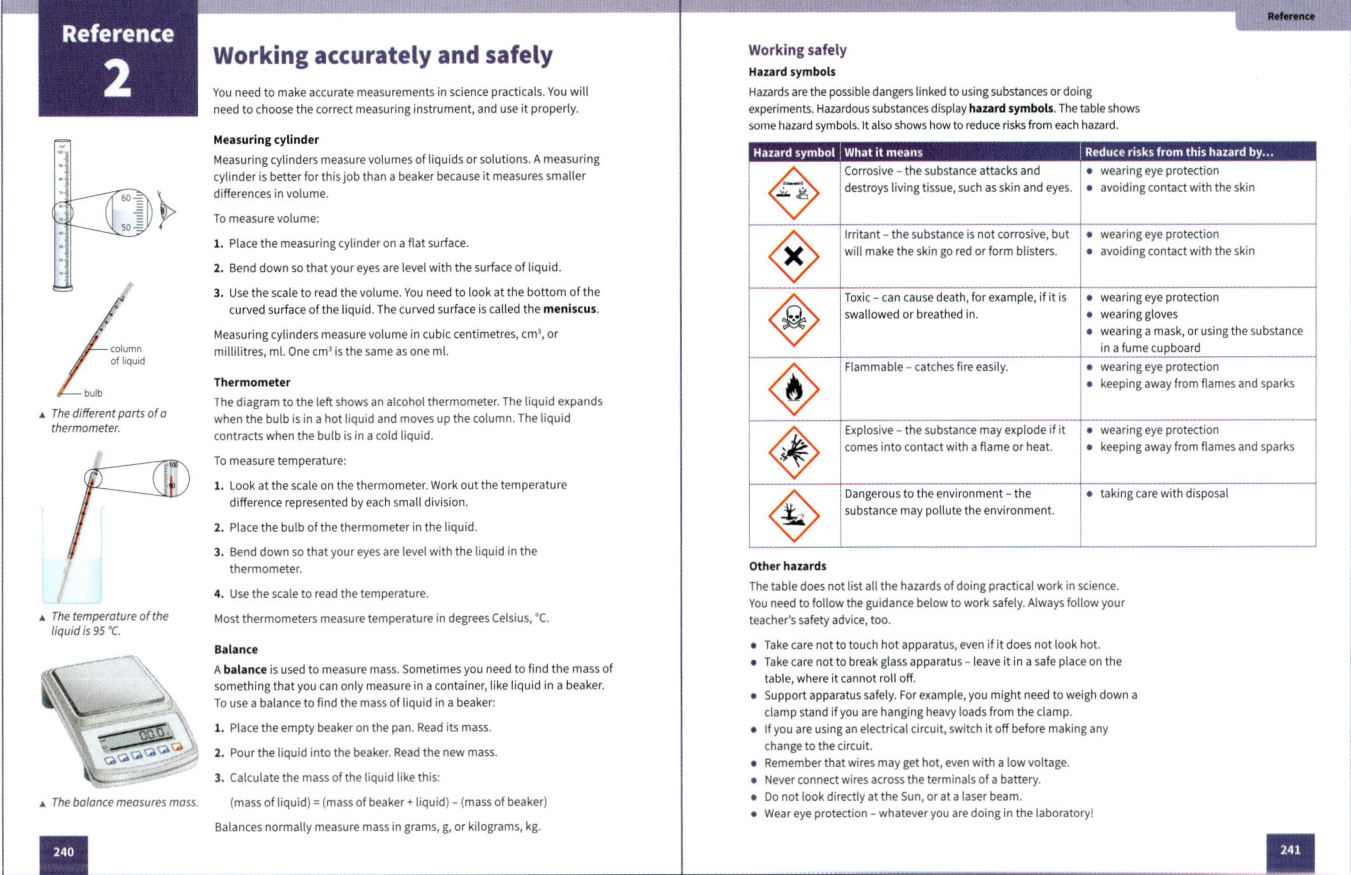

They include information on:

- how to choose suitable apparatus
- how to work accurately and safely.

There is also a periodic table of the elements at the end of this book, which you can quickly refer to whenever you need to.

Thinking and working scientifically

1

Objectives

- Recognise that there are many ways to find answers to questions in science
- Understand how to decide on a question to investigate
- Understand that there are some questions that science cannot answer

---

Part of making a prediction is to think about what might happen if your hypothesis is wrong. Your investigation should be able to show the difference between a correct and an incorrect hypothesis. Your conclusion will say whether the evidence supports, or does not support, your hypothesis.

▲ Subira's question can be answered by doing a fair test.

Asking questions

How do scientists answer questions?

We can ask lots of different questions about the world. Why does the battery last longer in some mobile phones than others? What might mobile phones be like in the future? Which mobile phone is best?

- There are questions that science can answer
- There are questions that science cannot answer.

What makes a question 'scientific'?

Scientists make **observations** and ask questions such as, 'How do fossil fuels form?' or 'Why are there are so many different animals on Earth?' These are scientific questions.

A scientific question is a question that you can answer by collecting and thinking about **data**. Data can be numbers from measurements, or words from observations.

Hypotheses and predictions

When they have a question, scientists may produce a **hypothesis.** A hypothesis is a scientific theory or proposed explanation made on the basis of evidence that can be further tested. A **prediction** is what you think will happen in the future. Scientists base their predictions on a hypothesis. Then they do an investigation or make further observations to collect data to see if their prediction is correct.

- A hypothesis is **testable** if you can:
- write a prediction based on the hypothesis
- collect data to see whether your prediction is correct.

Types of investigation

Scientists do **investigations** to collect data. There are lots of different types of investigation, for example:

- a fair test
- making a model
- a field study
- a survey or set of observations over time.

Fair testing

In science, anything that might change during an experiment is called a **variable**. The thing that you deliberately change to see whether it affects the outcome of the experiment is a variable. Anything that is affected as a result of your change is also a variable.

In some situations, scientists design an experiment to try to answer their question. To be sure of the answer, they must make it a **fair test**. In a fair test, the scientists change one **variable** to find out what effect it has, and they are careful to keep all the other variables the same.

The quantity that you change is the **independent variable**. A quantity that changes as a result is called a **dependent variable**.

Making a model

Sometimes it is not possible to do a practical investigation to answer a question – maybe what you are looking at is too big or small or dangerous to experiment on. Scientists can also make **models**. As well as helping to answer the question, a model can also be used to predict or to explain. Two types of model are a **physical model** and a **computer model.**

- A physical model is useful for very large-scale or small-scale systems. You may have used a physical model of the Earth and the Sun to explain why we have day and night.
- A computer model uses a computer program to find answers.

Field study

A **field study** is an investigation into plants or animals in their natural habitat. When doing fieldwork, it is important that you make observations without affecting what you are looking at.

A survey or regular observations or measurements

To answer some questions, a scientist might make lots of observations or measurements, or do a **survey**. They might do this over a long time, or all at the same time but in different locations. Sometimes a scientist uses data that other scientists have collected before.

Questions that science can't answer

Scientists cannot answer every question. They cannot answer questions about opinions, or questions for which the answer does not depend on data.

Science cannot answer Sanaa's question.
It could tell you:

- which phone battery lasts longest
- which phone can access web pages fastest.

But an investigation, a field study, observations, or a model will not tell you which phone is best. This is because different people will have different opinions about what is important: some people want a big screen, some people want a good camera, some people want a tough case.

In a fair test, you change the independent variable, measure the dependent variable, and keep all the other variables the same. The other variables are called **control variables**.

▲ *Thulani's question can be answered by making lots of observations of chimpanzees in their natural habitat. She would collect data and then choose the best way to display it.*

▲ *Mosi's question could be answered by collecting data from lots of different countries and comparing them.*

▲ *Science cannot answer Sanaa's question.*

Thinking and working scientifically 2

Planning and carrying out investigations

How do scientists get the data they need to find the answer to a scientific question? They need to make a plan to collect accurate and precise data.

Planning fair tests

Before carrying out a fair test, you should make a plan. This helps to make sure that you have all the equipment you need to get useful results, and that you do not forget to do something, or do anything dangerous.

Selecting equipment

You need to select equipment that enables you to make measurements of your independent and dependent variables. You may also need equipment that will help you to control the other variables.

You should think about which equipment is most appropriate and how to use it appropriately. For example, you may need to decide whether a measuring cylinder or a beaker is better for measuring volume, or how to measure length accurately.

Accurate and precise data

The measurements you make in an investigation are called data.

It is important to collect data that is **accurate** and **precise**.

Accurate data are close to the true value of what you are trying to measure.

Precise data give similar results if you repeat the measurements. The repeat measurements in each set are grouped closely together. Precision is also determined by the smallest division of the measuring instrument you are using.

Reliability

You need to be confident that your data is **reliable** when you make a conclusion. Data is reliable if you have taken enough measurements. How many are enough?

- You need to have a big enough **range** of values of the independent variable. The range is the difference between the biggest and smallest values. If your range is not large enough you may realise after the investigation that you cannot be confident in your conclusion.
- You need to repeat your measurements. We usually make three **repeat measurements** for every value of the independent variable. You can do fewer, or more.
- You need to deal with **anomalous results**. An anomalous result is a measurement that is very different from the others in a set of repeat measurements and might be a mistake.

Objectives

- Describe how to plan a fair test
- Describe how to plan other types of investigation

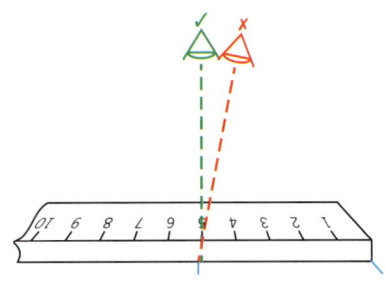

▲ You should look straight at a scale to make an accurate measurement.

not accurate
not precise — accurate
not precise

not accurate
precise — accurate
precise

▲ Readings can be precise but still not accurate.

Risk assessments

A plan should also include a **risk assessment**. This explains how you will reduce the chances of:

- damage to equipment
- injury to people.

Risk depends on the probability of the damage or injury happening, and the consequence if it did. You can reduce risk by:

- reducing the probability of something going wrong (e.g. keeping glass objects away from the edge of the desk)
- reducing the consequence if something goes wrong (e.g. wearing safety goggles).

Sometimes risks appear very small. You should still note them and say that they are not significant when you write your plan.

You should know the meaning of any hazard symbols, and consider them when you are planning your investigation.

▲ *Hazard symbols for flammable (left) and corrosive substances (right).*

What should a plan include?

Your plan should include:

- the scientific question that you are trying to answer
- why you have chosen to do a fair test
- the independent and dependent variables
- a list of variables to control, how you will do that, and the values each control variable will have
- your hypothesis: the scientific reason on which to base your conclusion
- your prediction: what you think will happen
- the list of the equipment you will need
- how you will use the equipment, step by step, to collect accurate and precise data
- how you will record your results
- a risk assessment, even if the risk is very low.

Planning other investigations

Not all investigations are fair tests. Some questions are better answered using a field study, making observations, or using **secondary data.**

Secondary data have been collected by other people. If you are planning to use secondary data you need to be confident that:

- the information is reliable
- the observations are accurate.

What should a plan for other investigations include?

There are some similarities in the plans for fair tests and other investigations. You should include:

- the scientific question that you are trying to answer
- why you are using this type of investigation
- how you are collecting precise and reliable data or observations
- the sources of secondary information, if you are using it
- your hypothesis: the scientific reason on which to base your conclusion
- your prediction: what you think your data will show
- the list of the equipment you will need, if you are doing it yourself
- a risk assessment, even if the risk is very low.

Thinking and working scientifically 3

Collecting and recording data

How do scientists collect and record the data that they need to answer scientific questions?

Objectives

- Describe how to record data from a range of investigations
- Describe how to deal with anomalous results
- Describe how to calculate the mean (average)

Using tables

Measurements are easier to understand if they are in a clear table.

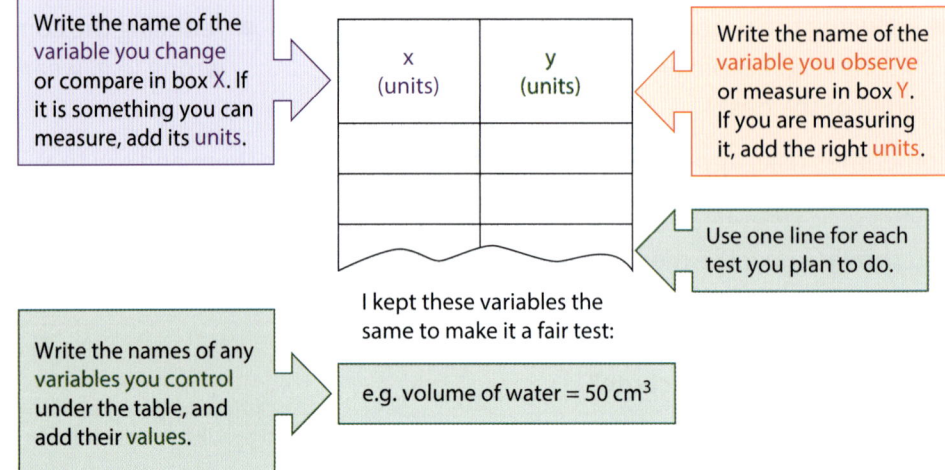

Using the correct units

If you use the wrong units for your measurements, your calculations will be wrong.

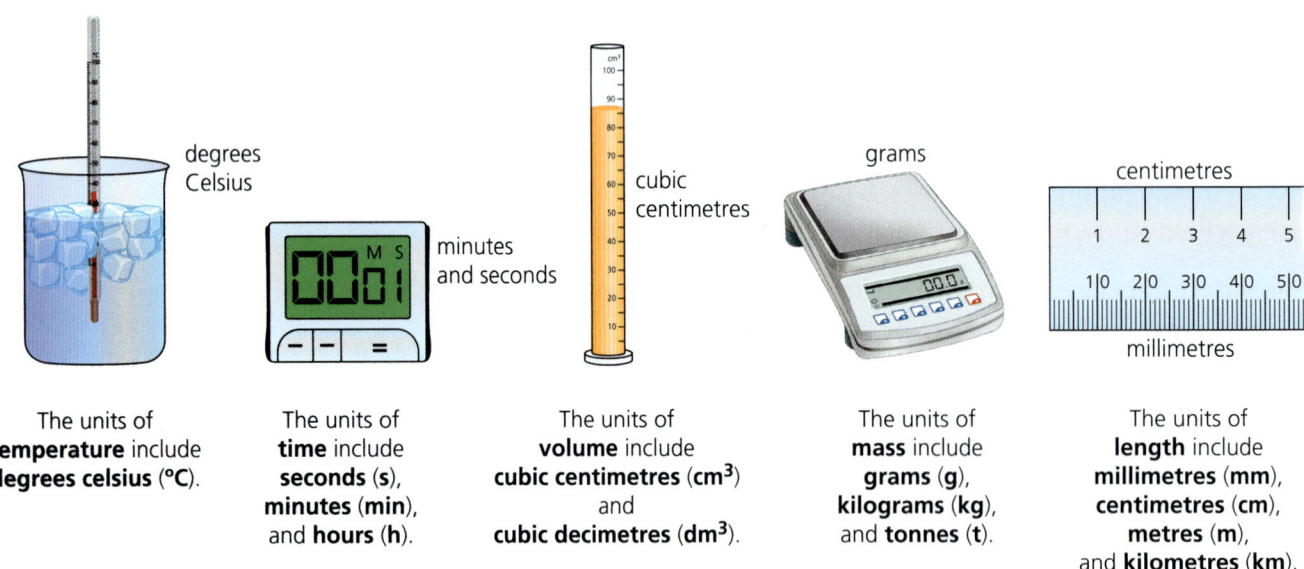

The units of **temperature** include **degrees celsius (°C)**.

The units of **time** include **seconds (s)**, **minutes (min)**, and **hours (h)**.

The units of **volume** include **cubic centimetres (cm³)** and **cubic decimetres (dm³)**.

The units of **mass** include **grams (g)**, **kilograms (kg)**, and **tonnes (t)**.

The units of **length** include **millimetres (mm)**, **centimetres (cm)**, **metres (m)**, and **kilometres (km)**.

Recording repeat measurements and calculating the mean

In most fair tests you should find the value of the dependent variable more than once, and usually three times. This is a **repeat measurement**. You should record all your repeat measurements in your results table, always to the same number of decimal places.

When you have finished the repeated measurements, you should:

- Check for any **anomalous results**. Do not erase them.
- You can repeat the measurement, and if one is very different from the other two put a line through it and ignore it. Use your new measurements.
- When you are confident that you do not have any anomalous results, calculate the average (**mean**) of the measurements.
- To calculate the mean you add up the all the repeats and divide by the number of repeats.

For example, three students find the time it takes to draw a table. Jamil takes 75 seconds, Abiola takes 35 seconds, and Karis takes 73 seconds.

Abiola's result is anomalous because it is very different from the others. Jamil and Karis find out why. Abiola's table is very messy. She did not use a ruler. They decide to leave it out of the mean. The mean is (75 s +73 s)/2 = 74 s.

Your average should be rounded up to the same number of decimal places as in the data.

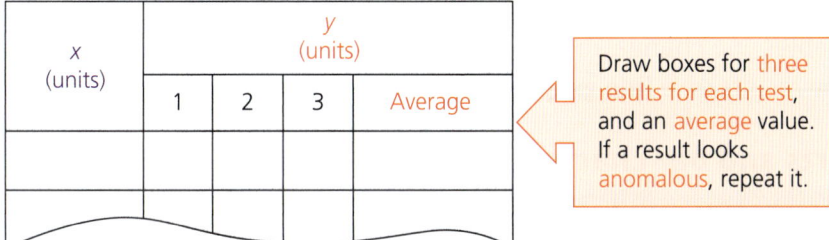

Draw boxes for three results for each test, and an average value. If a result looks anomalous, repeat it.

Recording measurements and calculated quantities

There are some experiments where you will need to calculate quantities using the measurements you have made.

In this situation, you should make another column in your table.

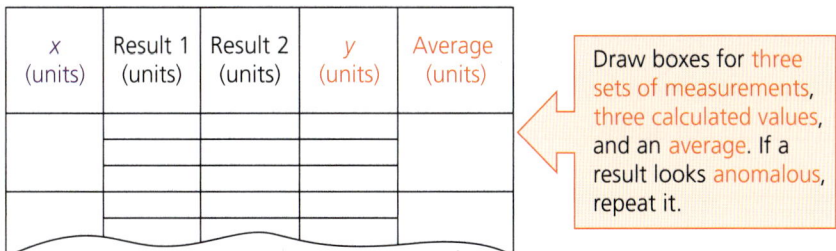

Draw boxes for three sets of measurements, three calculated values, and an average. If a result looks anomalous, repeat it.

Recording observations

In investigations where you are recording your observations, you need to adapt your table. In some cases, you may want to include images or diagrams, so the boxes in your table should be large enough to include them.

Thinking and working scientifically 4

Objectives

- Describe how to decide which graph to plot
- Describe how to draw a bar chart, a line graph and a scatter graph
- Describe how to draw a line of best fit

Drawing graphs

How do you know which type of graph to plot? What is the best way to plot graphs? The way that you display the results of your investigation depends on the type of data that you have collected.

Types of data

If the values of the variable you change (x) are *words*, then x is a **categoric** variable. There is no logical order, like size, for the categories. Names are one example.

Variables like shoe size are **discrete** variables. They are numbers, but there are no in-between sizes. The number of paper clips in a pot or people in a room are discrete variables.

You can only draw a **bar chart** or a **pie chart** for data that include categoric or discrete variables.

Other variables are **continuous** variables. Their values can be any number. Height, temperature, and time are continuous variables.

If the variables you change and measure are *both* continuous variables, display the results on a **line graph** or **scatter graph**.

Pie charts and bar charts

Student	Time spent on poster (minutes)	Time spent on homework (hours)
Deepak	24	2.5
Jamila	54	4.5
Kasim	12	2.0

Deepak, Jamila, and Kasim research and design a poster together.

The table shows the time that each of them spends. The pie chart drawn from the results helps you to see who did the most work on their project.

Pie charts are useful for showing fractions of a whole. When you want to show data that do not add together, a bar chart is better.

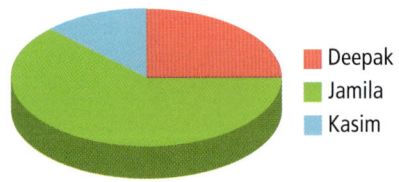

▲ *Time spent on homework.*

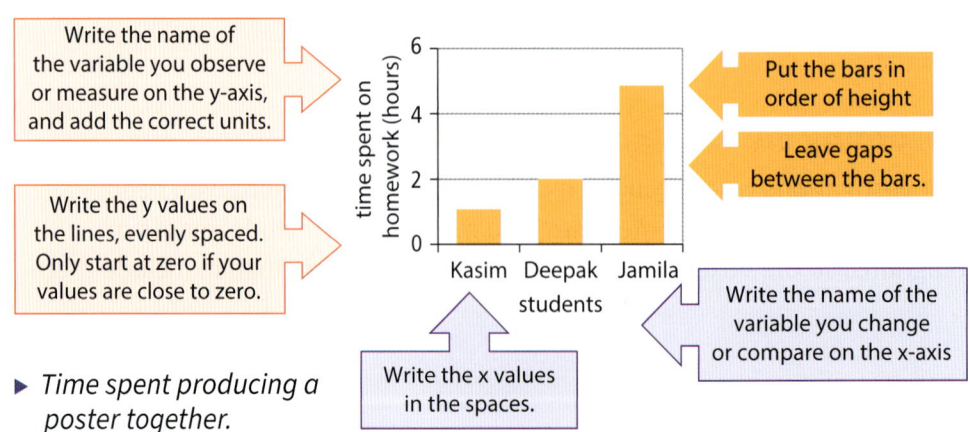

▶ *Time spent producing a poster together.*

Line graphs and scatter graphs

Drawing a line graph

A line graph makes it easier to see the link between two continuous variables – the **independent variable** and the **dependent variable**.

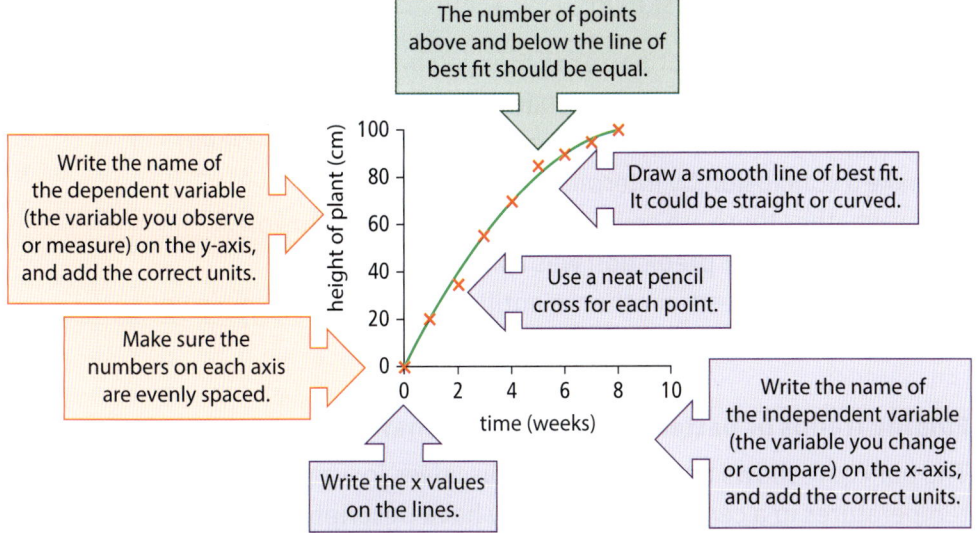

▲ Use a line graph for continuous variables when you think there is a link between them.

Drawing a scatter graph

A scatter graph shows whether there is a **correlation** between two continuous variables. In the graph below, all the points lie close to a straight line. That means there is a correlation between them. If there is no correlation between the variables, then the points would be scattered all over the graph.

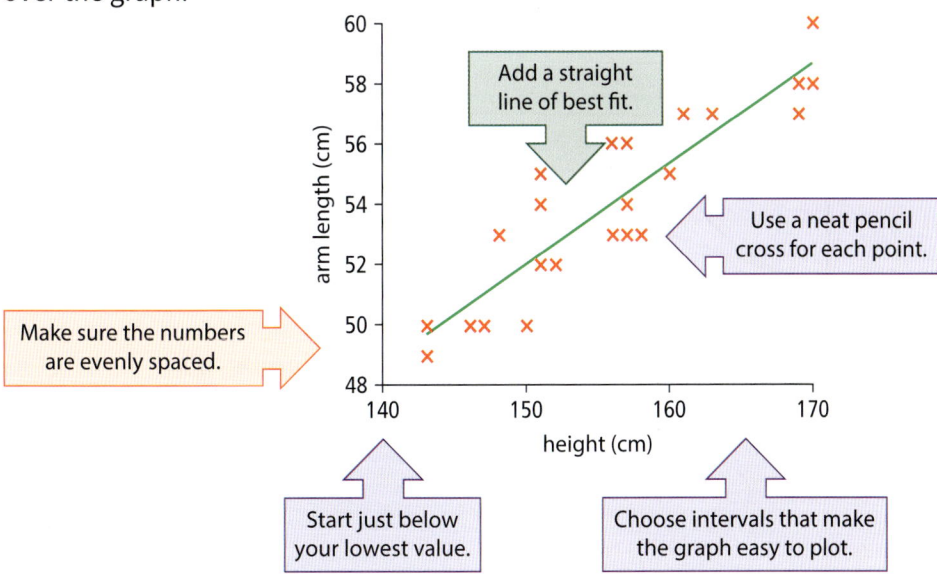

▲ A scatter graph will show you if there is a correlation between two continuous variables.

If you collect continuous data in a fair test investigation, you are trying to find out how one variable affects the other. You will usually plot a line graph.

In other investigations, you may be trying to see if there is a relationship between two variables. You will usually plot a scatter graph.

A line graph shows the link between two variables. You should draw a **line of best fit**. This is a line that goes through as many points as possible with roughly equal numbers of points either side of the line.

A correlation does not mean that one variable affects the other one. Something else could make them both increase or decrease at the same time. For example, if you plotted the number of ice creams sold in a town each day against the number of people going to the town swimming pool that day, you would see a correlation. This does not mean that getting wet makes people eat ice cream, or that eating ice cream makes people go swimming. It probably means that on hot days more people want to go swimming and to eat ice cream.

Thinking and working scientifically

5

Objectives

- Describe how to do an analysis of an investigation
- Describe the relationship shown by different lines of best fit on graphs

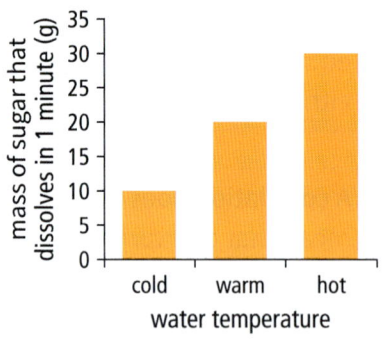

▲ Mass of sugar which dissolves in water at different temperatures.

Analysis

Analysing the evidence

When you analyse the evidence that you have collected (yourself or from secondary sources) you should:

- describe the **trends** or **patterns** that you have worked out from the display of your data (a graph, chart, or other display)
- identify any anomalous results, and suggest reasons for them
- make a conclusion by interpreting the results
- say whether there are any limitations to your conclusion
- say whether your prediction was correct
- use your hypothesis or other scientific knowledge to explain your conclusion.

Finding trends or patterns in graphs and charts

The bar chart shows how much sugar dissolves in water at different temperatures in a certain time. We only have descriptions of the temperature, not numbers, so the results are categoric.

You can describe the trend by saying:

'As the temperature of the water increases, the mass of sugar that dissolves increases.'

This is sometimes called the **relationship** between the variables.

Line or scatter graphs show relationships between continuous variables. When you have plotted the points on a line or scatter graph, draw a line of best fit.

In the graphs below the line of best fit is shown, but not the points.

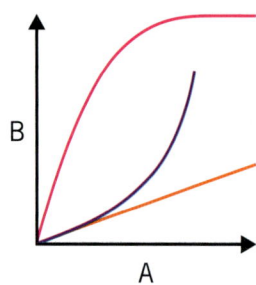

▲ In these graphs, if A increases then B increases

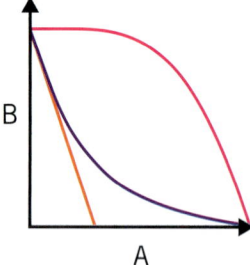

▲ In these graphs, if A increases then B decreases

Line or scatter graphs will be different if there is no relationship or correlation between the variables.

You may get a horizontal line in a fair test investigation if changing the independent variable has no effect on the dependent variable.

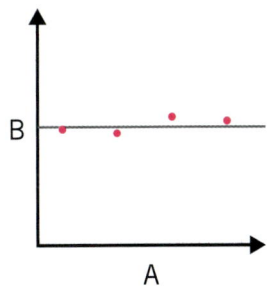

▲ *In this graph, if A increases B does not change.*

Identifying anomalous results

When you recorded your data, you may have identified points that did not fit the pattern. You may have ignored them, or repeated the measurement.

- You should identify these points now.
- You might see points that are not close to your line of best fit. They too are anomalous results.
- You should think of possible reasons why they might have occurred.

Writing your conclusion

A conclusion states what you have found out. You should also think about the **limitations** to your conclusion. Your conclusion may be limited if:

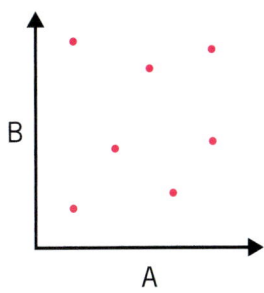

▲ *In this graph, there is no correlation between A and B.*

- you had lots of anomalous results
- the line of best fit is not clear
- your data is limited in terms of the range of variables that you investigated
- your data was limited in terms of the number of results that you collected.

> If points are scattered everywhere on a graph of two variables, it shows that they do not affect each other.

Checking your prediction

When you have found the pattern, you need to check the prediction you made and say whether it was correct. You should look carefully at the extent to which the evidence (data and observations) supports or refutes (disproves) your prediction.

Explaining your conclusion

Finally, *suggest scientific reasons* for any relationship or correlation differences that you have found. You could refer back to your hypothesis, or other scientific knowledge.

I think that the powdered sugar dissolved faster than the normal sugar because the pieces were much smaller.

I think that the shoe on the carpet needed a bigger force to move it than the shoe on the wooden floor because there was more friction between the shoe and the carpet.

I think that the plant near the window grew more quickly than the plant away from the window because there was more sunlight there.

Thinking and working scientifically 6

Objectives

- Describe how to evaluate data that you have collected
- Describe how to evaluate methods that you have used
- Describe improvements that you can make to improve the quality of the data

▲ *Evaluating means working out what is good and what is not so good.*

Evaluation

After you have collected your data, plotted a graph, written a conclusion, and explained what happened using scientific knowledge, you need to evaluate your investigation.

An evaluation is done in three stages:

- evaluate the *quality of the data*
- evaluate the *methods you used*
- *suggest improvements* to the method.

If you were to do the investigation again, the improvements should make you more **confident** in your conclusion.

Evaluating the data

Are there anomalous results?

When you look at your data tables and graphs, you can see how many anomalous results you had. This is why you do not erase the anomalous results in your results tables.

Anomalous results can limit your conclusions. They can also reduce the confidence that you have in your conclusion.

What is the spread?

The **spread** is between the smallest and the biggest values of repeated measurements. When you repeated the experiment, were the results close together (small spread) or far apart (large spread)? A smaller spread means that you can have more confidence in any conclusion based on your data.

What is the range and number of values?

The **range** of the variables that you have investigated is the difference between the smallest and the largest values. If the range is large, then you can be more confident of your conclusion.

You should collect enough data points to feel confident that you have correctly identified the trend or pattern. Two would not be enough.

Were there systematic or random errors?

There is **uncertainty** in any measurement that you make. This is one of the reasons why there is usually a spread in experimental data.

You should think about possible errors, as well as any anomalous results and the spread, to help you to decide how confident you are in your conclusion. There are two types of error that can affect scientific measurements.

- **Random errors** – these can affect the spread, or cause anomalous results. An example is the temperature of the room suddenly changing because someone opens a door.
- **Systematic errors** – these can make your measurements less accurate. An example is a newtonmeter reading 1 N even when there is nothing attached to it.

Evaluating the methods

When you planned your investigation, you chose the measuring instruments and the methods of using them to collect your data. You should look back at those choices and describe:

- the extent to which the *equipment* enabled you to collect data that was accurate and precise
- the extent to which the *methods* enabled you to collect data that was accurate and precise

I don't think I took the measurements of the volume very accurately because I did not look directly at the scale.

Suggesting improvements

Suggesting improvements is not about making the experiment easier or quicker.

Any improvements that you suggest should be designed to improve the *quality of your data*, because that would mean:

- there are fewer limitations to your conclusion
- you would be more confident in your conclusion.

I think we could have videoed the falling object so we could have played the video to see where it was.

How do I get better data with the same equipment?

You may improve the quality of the data by:

- eliminating systematic errors
- reducing the effect of any random errors
- including a bigger range
- doing more repeat readings
- repeating any measurements that gave anomalous results.

What other equipment would help?

Some types of equipment produce more precise and accurate data than others. You may need to do some research to find out.

- Using light gates to measure speed produces more accurate measurements than clicking a stopwatch by hand.
- Using a measuring cylinder to measure volume produces more accurate measurements than using a beaker.
- Using a balance that measures to more decimal places produces a more precise measurement.

1.1 The particle model

Objectives

- Define the terms *material, substance, particle model, property*
- Describe the particle model
- List the factors that give a substance its properties

The picture shows a lake in Sikkim, India. Imagine visiting the lake and the mountains. What are they made from?

What is matter made from?

The different types of matter (stuff) that things are made from are **materials**. There are millions of materials. The water of the lake is a material. The rock of the mountains is a material. The gases of the air are materials, too.

All materials are made up of tiny **particles**. Particles are too small to see. One grain of sugar is made up of about 1 000 000 000 000 000 000 particles.

Substance or material?

Most materials are mixtures. Wood is a mixture. So is soil and most rock. But some materials are not mixtures. They have just one type of matter. A material that has one type of matter is called a **substance**. Substances include silver, sugar, and pure water.

All the particles in a substance are the same as each other. One water particle is the same as all other water particles. One silver particle is the same as all other silver particles.

But the particles of different substances are different. Water particles are different to silver particles. Silver particles are different to sugar particles. Every substance has its own type of particle.

What are properties?

Every substance has its own properties. The **properties** of a substance describe what it is like and what it does. The pictures show some of the properties of three substances at room temperature (20 °C).

▲ Silver is shiny. Its shape stays the same unless you apply a force.

▲ Water has no colour. It flows.

▲ Sugar is white. It has a sweet taste.

The particle model

How does the particle model explain properties?

The **particle model** describes the arrangement and movement of the particles in a substance. You can use the particle model to explain properties. The properties of a substance depend on five factors:

- what its particles are like
- how its particles are arranged
- how its particles move
- how far apart the particles are (their separation)
- how strongly its particles hold together (their attraction for each other).

The pictures and captions show how the particle model explains some properties.

▲ A gold particle is a similar size to a copper particle. But a gold particle is heavier than a copper particle. This explains why a gold coin is heavier than a copper coin of the same size.

▲ Liquid water and steam have the same particles. But the particles in steam are further apart than the particles in liquid water. So, one gram of steam takes up more space than one gram of liquid water.

▲ The particles in liquid water move around. The particles in solid rock are held in one place. This explains why liquid water flows but solid rock does not flow.

▲ The particles in silver hold together less strongly than the particles in diamond. This explains why silver is easier to scratch than diamond.

Questions

1. Write definitions for *substance, material, property, particle model*.
2. Describe three properties of silver at room temperature.
3. Choose the one factor from the list that explains why ice is easier to scratch than silver.
   List of factors: what the particles are like, how the particles are arranged, how strongly the particles hold together.
4. A gold particle is the same size as a silver particle. Suggest why a gold ring is heavier than a silver ring of the same size.

📖 Key points

- Matter is made up of particles.
- A substance has one type of particle.
- The properties of a substance describe what it is like and what it does.
- The particle model explains properties.

1.2 The states of matter

What do the pictures have in common?

Objectives

- Define the term *physical property*
- List the three states of matter
- Describe differences in the physical properties of a substance in its three states
- Describe the arrangement, movement, separation, and attraction of the particles in the three states

The pictures show water in its three states: solid, liquid, and gas. Most substances can exist as a solid, liquid, or gas. These are the three **states of matter**.

What are the properties of a substance in its three states?

Ritula has a jug of liquid water. She pours some water into a glass. The shape of the water changes.

Ritula also has some ice shapes. Ice is water in the solid state. If Ritula keeps the ice cold, the shapes do not change.

Shape – and whether or not it can change – is one property that depends on the state of a substance. The table shows some other properties that depend on state.

▲ *Liquid water changes shape.*

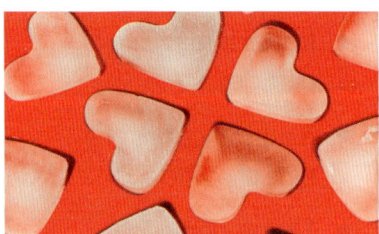

▲ *Solid water (ice) shapes do not change shape.*

State	In this state, does a substance flow?	In this state, can you compress (squash) a substance?	In this state, can the shape change?
solid	no	only a tiny bit	shape stays the same unless you apply a force
liquid	yes	only a tiny bit	takes the shape of the bottom of its container
gas	yes	yes, a lot	takes the shape of the whole container

The properties in the table are physical properties. **Physical properties** are properties that you can observe or measure without changing a material.

What do the particles do in the three states?

The particle model explains the physical properties of a substance in its three states. The particles themselves are the same in all three states. For example, a particle in ice is the same as a particle in liquid water and in steam.

But some factors are different for a substance in the three states:

- how its particles are arranged
- how its particles move
- how far apart the particles are (their separation)
- how strongly the particles hold together.

The solid state

In the solid state, the particles touch each other. The particles are in fixed positions in a regular pattern. The particles vibrate on the spot. The particles hold together strongly.

The liquid state

In the liquid state, the particles touch each other. They are not in a pattern. The particles move around randomly, sliding over each other. The particles hold together strongly.

The gas state

In the gas state, the particles are far apart. They are not in a pattern. The particles move around very fast, in all directions. The particles do not hold together strongly.

The table summarises the arrangement and behaviour of the particles in the three states.

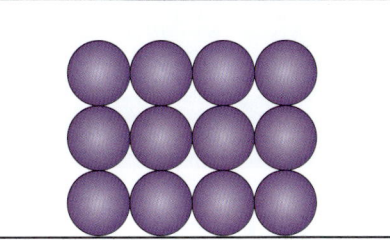

▲ Some particles of a substance in the solid state.

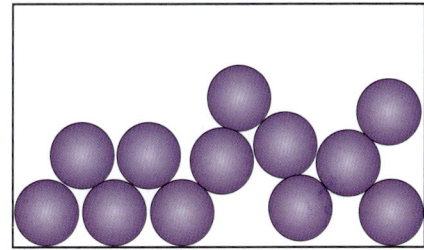

▲ Some particles of a substance in the liquid state.

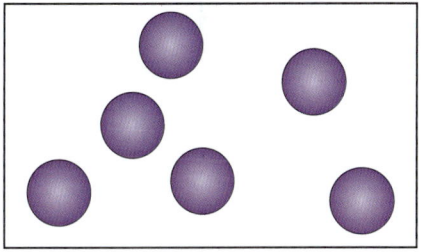

▲ Some particles of a substance in the gas state.

State	How close are the particles?	How are the particles arranged?	How do the particle move?	How strongly do the particles hold together?
solid	touching	in a pattern	vibrate on the spot	strongly
liquid	touching	randomly	move around randomly, sliding over each other	strongly
gas	far apart	randomly	move around randomly, in all directions	very weakly

Questions

1. List the three states of matter.
2. Describe two differences and one similarity between the physical properties of a substance in the solid and liquid states.
3. Describe the arrangement, movement, and separation of the particles in the liquid state.
4. Compare how strongly the particles hold together in the liquid and gas state.

Key points

- There are three states of matter: solid, liquid, and gas.
- Physical properties are those you can observe or measure without changing a material.
- A substance has different physical properties in its three states.
- The particles have different arrangements in the three states, and they move and are held together differently.

1.3 Using the particle model

Objectives

- Use the particle model to explain the physical properties of substances in their three states
- Define the term *vacuum*
- Describe one application of science

Science in context

Keeping vaccines cold

Solid carbon dioxide is also called dry ice. Dry ice can keep vaccines very cold. It is packed into boxes around the vaccine bottles.

▲ Dry ice keeps one COVID-19 vaccine colder than −70 °C.

The picture shows carbon dioxide in the solid state. The shape and size of each lump does not change, if it is kept cold. But in the gas state, carbon dioxide spreads out and mixes with the air, all around the Earth. Why is this property so different in the two states?

How does the particle model explain shape?

In solid carbon dioxide, the particles are arranged in fixed positions. They hold together strongly. This explains why the shape and size of each lump does not change.

In the gas state, the particles move around. They do not hold together strongly. This explains why carbon dioxide gas spreads out. It takes the shape of its container.

◀ Pouring water

This boy is playing with liquid water. He pours it from one jar to another. The shape of the water changes, because the particles move around.

How does the particle model explain other physical properties?

Different parts of the particle model explain different properties.

Flow

A substance can flow in its liquid and gas states. This is because the particles move around. But in the solid state, the particles are in fixed positions. They cannot move. This explains why, in the solid state, a substance cannot flow.

The particle model

▲ Liquid mercury flows, but solid mercury cannot flow.

▲ Carbon dioxide flows in the gas state. Solid carbon dioxide cannot flow.

Compression

There is enough oxygen squashed into the firefighter's cylinder to support her for several hours. Normally, this amount of oxygen fills 1000 soda bottles.

You can **compress** (squash) all substances in the gas state. This is because the particles are far apart. They get closer together when the gas is compressed. The particles themselves stay the same – they do not change size.

In the solid and liquid states, a substance can be compressed only a very tiny bit. This is because the particles are already touching.

What is a vacuum?

A space that has no particles (and so no matter) is called a **vacuum**. Most of outer space is close to being a vacuum. But there are a few particles in outer space. They are very, very, very far apart.

▲ The firefighter's cylinder is full of oxygen.

Questions

1. Write the definition for a *vacuum*.
2. Choose the one factor from the list that can be used to explain why a gas can be compressed.
   List of factors: how the particles move, how the particles are separated, what the particles are like.
3. Use the particle model to explain why a lump of a solid does not change its shape.
4. Use the particle model to explain why the shape of liquid water can change.
5. Describe how a substance flows, or does not flow, in the solid, liquid, and gas states. Use the particle model to explain your answer.

Key points

- The particle model explains properties such as whether a substance flows or can be compressed.
- Different parts of the particle model explain different properties.
- A vacuum is a space that has no matter in it.

1.4 Changes of state – evaporating, boiling, and condensing

Objectives

- Define the terms *boiling, evaporating,* and *condensing*
- Describe how the arrangement, movement, and separation of the particles change when a substance boils, evaporates, or condenses
- Make conclusions from data

Chahaya lights a candle. Some of the solid wax melts, forming liquid wax. Some of the liquid wax evaporates, forming wax in the gas state. The wax gas burns.

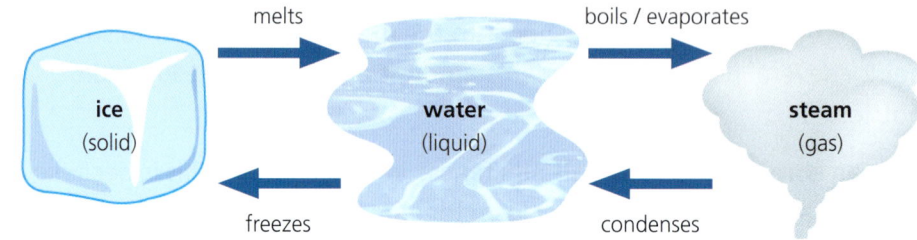

What are changes of state?

Melting and evaporating are **changes of state**. The diagram gives the names for six changes of state. You may already know the names.

ice (solid) — melts → water (liquid) — boils / evaporates → steam (gas)
steam — condenses → water — freezes → ice

Most substances change state, not just water.

How does the particle model explain changes of state?

When a substance changes from one state to another, its particles stay the same. They do not change. The things that change are:

- how the particles are arranged
- how the particles move
- the separation of the particles.

Evaporating or **evaporation** is the change of state from liquid to gas that can happen at any temperature.

In the liquid state, particles touch each other. They move around, sliding over each other. Some particles move faster than others. The faster-moving particles leave the surface of liquid. They separate from each other, forming gas. This is evaporation.

▲ Clothes dry when water evaporates from them.

▲ The eggs are cooking in boiling water.

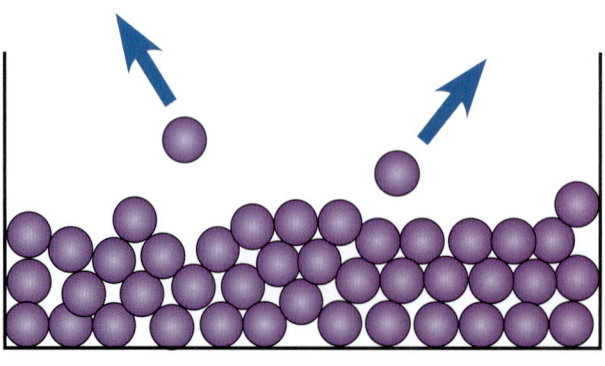

◄ During evaporation, particles leave the liquid surface. Not to scale.

Boiling is the change of state from liquid to gas. It only happens if the liquid is hot enough.

When you heat liquid water, bubbles of steam form everywhere in the liquid. In liquid water, the particles are touching each other. Inside the bubbles, the particles are spread out. The bubbles rise to the surface and escape. The water is boiling.

The temperature that a substance boils at is its **boiling point.** Different substances have different boiling points. A substance has a high boiling point if its particles hold together strongly. A substance has a low boiling point if its particles hold together weakly.

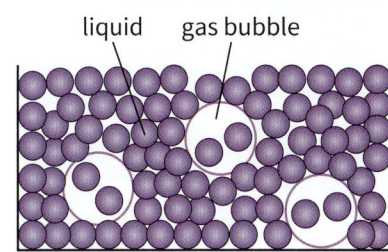

▲ When a liquid is boiling, bubbles of gas form throughout. Not to scale.

Process	Change of state	Temperature	How particles leave the liquid
evaporating	from liquid to gas	happens at any temperature	particles leave the liquid surface
boiling	from liquid to gas	happens at the boiling point	bubbles of the substance in the gas state form everywhere in the liquid; the bubbles rise to the surface and escape

▲ Table 1: A comparison of evaporating and boiling.

Thinking and working scientifically

Making conclusions from data

Table 2 shows the boiling points of six substances.

You can interpret the data in the table to make conclusions.

- First conclusion: Diamond has the highest boiling point. Its particles are held together most strongly.
- Second conclusion: The boiling point of nitrogen is lower than room temperature (20 °C). This means that nitrogen is in the gas state at room temperature.

Substance	Boiling point (°C)
nitrogen	−196
chlorine	−102
ethanol	78
water	100
mercury	357
copper	2595
diamond	5100

▲ Table 2

Condensing or **condensation** is the change of state from gas to liquid. A gas condenses at any temperature below its boiling point.

When a substance changes from the gas state to the liquid state, the particles in the gas move more slowly. They get closer until they are touching, forming liquid. The substance has condensed.

Key points

- A substance changes from liquid to gas by evaporating or boiling.
- A substance changes from gas to liquid by condensing.
- Every substance has its own boiling point.

Questions

1. Name the change of state from gas to liquid.
2. Describe the movement and separation of the particles before and after a substance evaporates.
3. Compare evaporating and boiling. Use Table 1 to help you.
4. Look at the data in Table 2. Name one substance, other than nitrogen, that is in the gas state at 20 °C.

Thinking and working scientifically

1.5

Objectives

- Describe how to collect evidence to test a hypothesis
- Draw a graph to present data
- Describe patterns in results and make conclusions

Investigating boiling points

Rabia is a school student. She lives in Mumbai, a city by the sea. She visits Leh, high in the mountains. She discovers that water boils at different temperatures in the two places. Rabia asks a scientific question:

Why does water boil at different temperatures in different places?

Planning the investigation

Making a hypothesis

Rabia makes a hypothesis. As you know, a hypothesis is a possible explanation that is based on evidence, and can be tested further.

My hypothesis – The boiling point of water depends on altitude (height above sea level). The higher the altitude, the lower the boiling point.

Rabia can collect evidence to find out if her hypothesis is correct. This means that the hypothesis is testable.

Carrying out the investigation

Rabia uses a thermometer to measure the boiling point of water in Mumbai. This is not enough evidence to test her hypothesis. She needs to collect more evidence. She asks her relatives to measure the boiling temperature of water in other places.

Some of her relatives send photos of the thermometers they use.

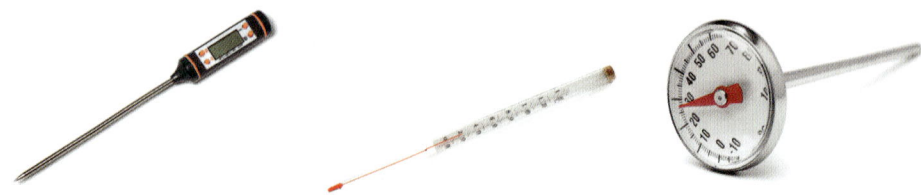

▲ *Different thermometers*

Rabia then uses the Internet to find the altitude of each place, and the boiling temperatures on two mountain summits. Rabia writes all the results in a table. This makes it easy to see all the data.

28

The particle model

Place	Altitude (metres above sea level)	Boiling temperature of water (°C)
Chennai, India	0	100
Jos, Nigeria	1200	96
Leh, India	3524	88
Mumbai, India	0	100
Riyadh, Saudi Arabia	610	40
Srinagar, India	1585	95
Summit of Mount Everest, Nepal	8550	69
Summit of Puncak Jaya, Indonesia	4884	83

Analysing the evidence

Making a conclusion

Rabia uses the data in her table to draw a line graph. This will make it easier to spot a pattern.

Rabia notices that the boiling point for Riyadh does not fit the pattern. It is anomalous. She messages her relative there. He said that he made a mistake, and measured the air temperature, not the boiling temperature of water. Rabia decides to ignore this result.

Rabia writes a conclusion:

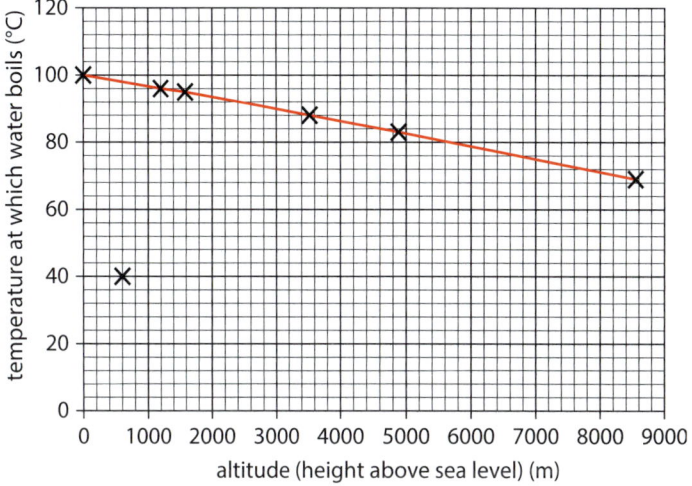

My conclusion – My results show that as altitude increases, boiling temperature decreases. This supports my hypothesis. A limitation of my investigation is that it does not explain **why** boiling temperatures are different at different altitudes.

Suggesting improvements

Rabia knows that her uncle in Srinagar boiled water in three different pots. He measured the boiling temperature in each pot. It was the same each time. Rabia knows that his data is precise. If she did the investigation again, she would ask everyone to repeat the experiment three times.

Questions

1. Write the definition for *hypothesis*.
2. Use the table to write down the boiling temperature of water at the summit of Mount Everest. Include units in your answer.
3. Describe the pattern shown by the graph.
4. The city of Addis Ababa, in Ethiopia, is at altitude 2400 m. Use the graph to predict the boiling temperature of water in Addis Ababa.
5. Look at the pictures of the thermometers. Suggest and explain an improvement that Rabia could make to the investigation, other than asking everyone to repeat the experiment three times.

Key points

- A hypothesis is a possible explanation that is based on evidence and can be tested further.
- An anomalous result is one that does not fit the pattern.
- In a conclusion, describe any patterns and explain any limitations.

1.6 Changes of state – melting and freezing

Objectives
- Define the terms *melting* and *freezing*
- Describe how the arrangement, movement, and separation of the particles change when a substance melts or freezes
- Make conclusions from data

Arin puts some butter in a pan. He heats the pan. As he heats, the butter starts to become runny. It is changing from the solid state to the liquid state.

How does the particle model explain melting?

The change of state from solid to liquid is **melting**. When a solid begins to melt, its particles vibrate faster. The particles move out of their fixed positions. They move around, sliding over each other. Some of the substance is now in the liquid state.

If the substance continues to be heated, more of its particles start to move around. Eventually, all the substance has melted. In the liquid, the particle arrangement changes all the time.

But melting does not change the separation of the particles. The particles touch each other in both the solid and liquid states.

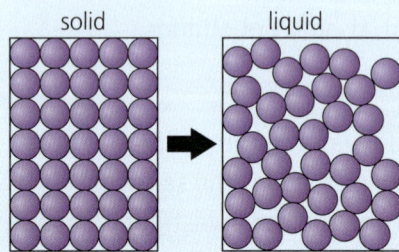

◀ On melting, particles leave their fixed positions.

Melting point

The temperature that a substance melts at is its **melting point**. Every substance has its own melting point. A substance has a high melting point if its particles are held together strongly.

▲ Gallium melts on your hand. Its melting point is 30 °C.

▲ This photo shows ice forming from liquid water.

Substance	Melting point (°C)
nitrogen	−210
water	0
gallium	30
paracetamol	169
gold	1063
copper	1083

30

Science in context

Identifying substances

Scientists use melting points to help identify substances. In a university laboratory, Gladys and Jamil follow instructions to make a painkiller, paracetamol. They end up with white crystals.

The students want to check that the crystals are definitely paracetamol. They put some crystals in a thin glass tube. They place the tube in the melting point apparatus. The substance melts at 169 °C. The melting point confirms that they have made paracetamol.

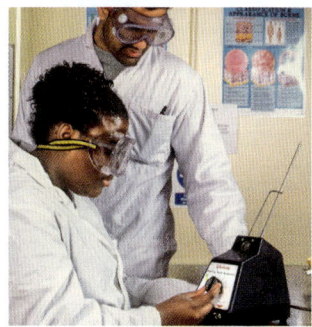
▲ *Using melting point apparatus.*

How does the particle model explain freezing?

The change of state from liquid to solid is **freezing**. A substance freezes at, or below, its melting point.

When a liquid freezes, its particles stop moving around from place to place. Over time, they arrange themselves in a pattern. Eventually, all the particles are in fixed positions. They vibrate on the spot. All of the substance is now frozen.

In melting and freezing, the particles themselves do not change. No particles are added or removed. This is true for all changes of state.

Thinking and working scientifically

Making conclusions from data: solid, liquid, or gas?

If you know the melting point and boiling point of a substance, you can predict its state at different temperatures:

- Below the melting point, it is in the solid state.
- Between its melting and boiling points, it is in the liquid state.
- Above its boiling point, it is in the gas state.

◀ *The diagram shows the melting point and boiling point of silver.*

Questions

1. Name the change of state from:

   a. solid to liquid b. liquid to solid c. liquid to gas.

2. Describe the movement, arrangement, and separation of particles before and after freezing.

3. Look at the diagram of the melting point and boiling point of silver. Give the state of silver at these temperatures:

   a. 700 °C b. 1400 °C c. 2400 °C d. 20 °C.

Key points

- Melting is the change of state from solid to liquid.
- Freezing is the change of state from liquid to solid.
- Every substance has its own melting point.

Thinking and working scientifically

1.7

Objectives

- Define the term *model* in science
- Consider strengths and limitations of the particle model

Models in science

Yana makes a model truck. How is the truck like a real truck? How is it different?

▲ A model truck   ▲ A real truck

The model truck is the same colour as the real truck. It is a similar shape. But there are differences between the real truck and the model truck. The model truck is smaller. It cannot move by itself, or carry heavy loads.

What are models in science?

In science, a **model** is an idea that explains observations and helps in making predictions. The particle model is one model. You can imagine it in your head. You can draw it. Or you can use something else to represent the particles, like peas.

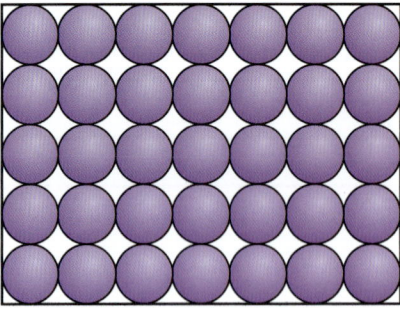 

▲ Two ways of representing the particle model – as a drawing, and with dried peas.

There are many other models in science. Some models represent things that are very small, such as particles and cells. Some models represent things that are very big, such as the Solar System. You can read about other models later in the book.

The particle model

What are the strengths and limitations of the particle model?

Strengths

Models help to explain things. The particle model explains why liquid water takes the shape of a cup, and why solid water does not. The particle model also explains changes of state, such as melting and freezing.

Models help in making predictions. If you know that the particles hold together strongly in solid gold, you can predict that gold has a high melting point.

▲ Gold has a high melting point.

Limitations

Models are simpler than reality. This is what makes them useful. But because they are simpler, they are not exactly the same. A model can never be perfect. Every model has limitations.

In the particle model, every particle is a solid sphere. In reality, different particles are different shapes. And in reality, particles are not solid. As you will see in Chapter 7, particles are mostly empty space.

▲ In the particle model, we imagine that all particles are spheres. But the shape of a water molecule is more like the shape on the right.

The differences between the particle model and reality give the model some limitations, or weaknesses:

- The particle model does not explain all observations perfectly.
- Predictions made with the particle model are sometimes not accurate.

Even though the particle model has limitations, it is still useful.

Questions

1. Write the definition for a *model* in science.
2. Use the particle model to explain why you can pour a substance when it is liquid, but not solid.
3. Describe two ways that the particle model is different from reality.
4. Describe two limitations of the particle model.

Key points

- In science, a model is an idea that explains observations and helps in making predictions.
- Every model has strengths and limitations.

33

Review 1.8

1.

   The picture shows water in two states.

   a. Name the two states shown in the picture. [2]

   b. Give the name of the state in which water takes the shape of the bottom of its container. [1]

   c. Give the names of two states in which water can flow. [1]

2. Choose one word from the list below to answer each question part.

properties	particle model
substance	physical properties
vacuum	evaporation

   a. These are properties that you can observe or measure without changing a material. [1]

   b. A space with no particles in it. [1]

   c. A material that has one type of matter. [1]

   d. These describe what a substance is like and what it does. [1]

   e. This describes the arrangement, separation, and movement of the particles in a substance. [1]

   f. The change of state from liquid to gas that can happen at any temperature. [1]

3. In the diagram below, each arrow represents a change of state.

   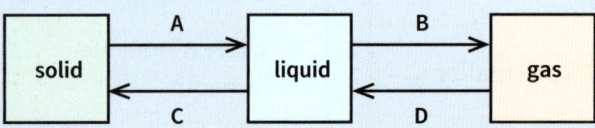

   a. Give the letter of the arrow that shows freezing. [1]

   b. Give the letter of the arrow that shows condensing. [1]

   c. Give the name of the change of state represented by arrow A. [1]

   d. Give the name of the change of state represented by arrow B. [1]

4. Use the words and phrases to copy and complete the sentences below. You may use each word or phrase once, more than once, or not at all.

a little	close together
far apart	gas
liquid	much
solid	vibrate on the spot
move around from place to place	
move around and slide over each other	

   Copper exists in three states – solid, liquid, and _____. In the solid state, its particles _____. The particles are _____. When copper melts, it changes state from _____ to _____. Its particles start to _____. They get _____ further apart. If copper is heated to 1084 °C, it changes from the liquid to the _____ state. Its particles get _____ further apart and they start to _____. [10]

5. The table gives the melting points and boiling points for five substances.

Substance	Melting point (°C)	Boiling point (°C)
bromine	−7	59
chlorine	−102	−34
iodine	114	184
osmium	3000	5000
tungsten	3422	5555

   a. Name the substance in the table with the highest melting point. [1]

   b. Name the substance in the table with the lowest boiling point. [1]

   c. Name one substance in the table that is in the gas state at 20 °C. [1]

d. Name two substances in the table that are in the solid state at 20 °C. [1]

e. Give the state of bromine at 20 °C. [1]

f. Copy and complete the sentence using words from the list. You may use them once, more than once, or not at all.
   solid   liquid   gas
   When bromine is heated from 20 °C to 100 °C it changes from the _____ to the _____ state. [2]

g. Name the change of state that occurs when osmium is heated from 4000 °C to 6000 °C. [1]

6. Write the letter of each label next to the correct line on the diagram. [3]

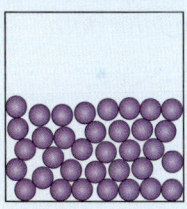

   A Water in the liquid state.
   B Water in the gas state (steam or vapour).
   C Mixture of air and steam.

7. Complete the sentences using words and phrases from the list. You may use each once, more than once, or not at all.
   increases   decreases   stays the same

   a. When liquid water boils, the separation of the particles _____ [1]

   b. When steam condenses, the speed of movement of the particles _____ [1]

   c. When steam condenses, the distance between the particles _____ [1]

8. Read the statements below about the particles in liquid water. All the statements are true.
   X The particles are close together in the solid and liquid states.
   Y The particles are not arranged in a regular pattern.
   Z The particles move around, sliding over each other.

   a. Write the letter of the one statement above that best explains why you can pour liquid water. [1]

   b. Write the letter of the one statement above that best helps to explain why the volume of 1 g liquid water is similar to the volume of 1 g of solid water (ice). [1]

   c. Write the letter of the one statement above that best explains why a liquid takes the shape of the bottom of its container. [1]

9. Oxygen is a gas at 20 °C.

   a. Describe the arrangement, separation, and movement of particles in the gas state. [3]

   b. Use the particle model to explain why oxygen gas can be compressed. [1]

10. The diagram shows the particles of a substance in its liquid state.

    a. Describe what you could do to make the particles in the liquid move more quickly. [1]

    b. Draw a diagram to show the same particles after some of the liquid has evaporated. [1]

11. Copy these and draw lines to match each property with the best explanation. [3]

Property	Explanation
You cannot compress a solid.	The particles move around, sliding over each other.
If a gas is in a container with no lid, it escapes from the container.	There is no empty space between the particles.
A liquid takes the shape of the bottom of its container.	Its particles are in fixed positions.
You cannot pour a solid.	The particles move around in all directions.

2.1 Elements and the periodic table

Two students are discussing substances. Who is correct?

All substances are materials, but not all materials are substances.

Materials and substances are the same thing.

Objectives

- Define the term *periodic table*
- Give examples of elements and their uses
- Find metal and non-metal elements in the periodic table

What are elements?

As you know, there are millions of materials. Most materials are mixtures. But some materials have just one type of particle. These are substances.

There are over 100 special substances called elements. An **element** is a substance that cannot be split into other substances. Each element has its own type of particle, and its own properties. Every material on Earth – and everything in the Universe – is made from the particles of one or more elements.

The element platinum is expensive. In 2020, 1 kg of the element cost more than 33,000 US dollars. Why is platinum so valuable? Because its properties give it important uses.

▲ Platinum-containing hard drives store computer data.

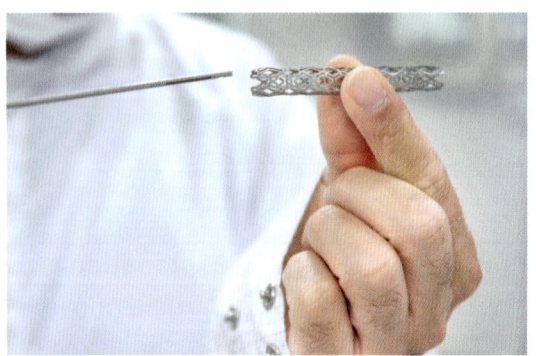

▲ A surgeon may insert a platinum-containing stent into a narrowed blood vessel, to keep it open.

▲ Catalytic converters in cars contain platinum. They convert harmful exhaust gases to less harmful ones.

▲ South Americans made platinum and gold jewellery like this about 2000 years ago.

What is the periodic table?

The periodic table lists all the elements in a certain order. It groups together elements with similar properties.

The periodic table opposite has a stepped line. This splits the elements into two important groups – metals and non-metals. The elements on the left of the line are **metals**. The elements on the right of the line are **non-metals**. You can read about the properties of metals and non-metals in Chapter 3.

36

Elements, compounds, and mixtures

Hydrogen is the most common element in the Universe. It is a non-metal.

All living things are mainly carbon (including you).

Nitrogen is the most common element in the Earth's atmosphere.

Helium is the second most common element in the Universe.

[Periodic table showing elements hydrogen through oganesson, with lanthanoids (lanthanum–lutetium) and actinoids (actinium–lawrencium) shown separately below.]

Many useful metal objects contain iron, which is a metal.

▲ A larger periodic table can be found at the back of this book.

Silver makes attractive jewellery.

📖 Key points

- An element is a substance that cannot be split into other substances.
- The periodic table lists all the elements, grouping together elements with similar properties.
- The uses of elements depend on their properties.

❓ Questions

1. Define the term *periodic table*.
2. Give one use of each of these elements: iron, platinum, silver.
3. Write the names of four metal elements and four non-metal elements. Use the periodic table to help you.

Science in context 2.2

Objective
- Describe some factors that influenced when elements were discovered

▲ This gold headdress was made over 4000 years ago. It was found in a grave in Ur, modern Iraq.

▲ Most early copper was extracted from malachite.

▲ These iron tools are from different parts of Europe.

Discovering the elements

Early ideas about elements

More than 2000 years ago, a Greek thinker suggested that there were four elements – air, water, fire, and earth. This idea was popular for hundreds of years. Scientists in India used the explanation to help develop their ideas about atoms.

From about 1700, scientists realised that the idea of four elements was wrong. They looked for substances that did not break down into simpler substances. They already knew about some of them…

The first discoveries

Some elements exist naturally on their own. Early humans saw sulfur near volcanoes. They made charcoal by burning wood. They found gold in stream beds.

▲ Sulfur on Ijen volcano in East Java, Indonesia.

▲ Charcoal is a type of carbon. It forms when wood burns.

Copper can also exist on its own, but most copper is found joined to other elements in rocks. People started extracting copper from its rock 7000 years ago. They used copper to make weapons, tools, and jewellery.

Later, people mixed copper with tin to make bronze. The properties of bronze mean that it is suitable for making knives.

People first extracted iron from rock about 3500 years ago. They made tools and weapons from the iron.

▲ The people of the Indus Valley started using bronze about 5000 years ago.

1200–1700 – zinc and phosphorus

It is not easy to get zinc from its rock. However, 800 years ago, scientists in India worked out how to obtain zinc by heating zinc-containing rock with other substances, including soot and nuts. They made tens of thousands of tonnes of zinc.

Hennig Brandt made the first phosphorus. He evaporated water from urine. He heated the solid that remained until it was red hot. He then collected the phosphorus vapour that evaporated.

The 1700s – elements from the air, and more metals

Scientists discovered 17 new elements in the 1700s. They isolated nickel from rocks and magnesium from pond water. They separated oxygen from the air.

Henry Cavendish and Joseph Priestley knew that air contains oxygen. They did an experiment to find out about the other gases in air. First, they removed oxygen from the air. Then they put a mouse in the remaining air. It died. Another scientist, Daniel Rutherford, explained the evidence. He said that air is mainly nitrogen.

The 1800s – technology drives discoveries

The 1880s saw the discovery of more than 50 elements. A new piece of equipment, the spectroscope, helped scientists to discover caesium and rubidium.

Scientists discovered aluminium in the 1800s. It was not found earlier because it is strongly joined to oxygen in its rock.

The twentieth and twenty-first centuries

The last gaps in the periodic table were filled in the 1920s. Ida Tacke and her colleagues discovered the elements rhenium and technetium.

Scientists made more than 25 artificial elements in the twentieth century.

▼ *The colours in this periodic table show when the elements were discovered.*

Note: This periodic table does not include all the elements.

Questions

1. Explain why people have known of sulfur, gold, and carbon for hundreds of years.
2. List five elements discovered between 1200 and 1700.
3. Name the two elements that were discovered in the 1900s.

Key points

- Elements that are found on their own were discovered first.
- Other elements were discovered as technology developed.

2.3 Chemical symbols

Objectives

- Write chemical symbols for elements
- Explain why scientists use chemical symbols

The pots in the picture below are glazed with a yellow powder that includes atoms of the element praseodymium. Chemists often write Pr instead of praseodymium. Can you suggest why?

What are chemical symbols?

Every element has its own **chemical symbol**. This is the one- or two-letter code for the element.

Often, the chemical symbol is the first one or two letters of the English name of the element.

Element	Chemical symbol
carbon	C
calcium	Ca
cobalt	Co
nitrogen	N
neon	Ne
nickel	Ni

Sometimes, the chemical symbol is made from the first and third letters of the English name of the element.

The chemical symbols for some elements come from other languages:

Element	Chemical symbol
magnesium	Mg
manganese	Mn
chromium	Cr
chlorine	Cl

- The chemical symbol for iron, Fe, comes from the Latin name for iron, *ferrum*.
- The chemical symbol for tungsten, W, comes from the German name for tungsten, *Wolfram*.

Scientists all over the world use the same chemical symbols. All scientists recognise S as the chemical symbol for sulfur, even though it is called *belerang* in Indonesian, كبريت (*kibriyt*) in Arabic, *soufre* in French, and *azufre* in Spanish.

Writing chemical symbols

Follow these rules to write chemical symbols correctly:

- Write a capital letter for a one-letter symbol. For example, the chemical symbol for nitrogen is N, not n.
- Write a capital letter followed by a lowercase letter for a two-letter symbol. For example, the chemical symbol for magnesium is Mg, not mg, MG, or mG.

Chemical symbols in the periodic table

As you know, the periodic table lists all the elements in a certain order. Most versions of the periodic table give element names and chemical symbols.

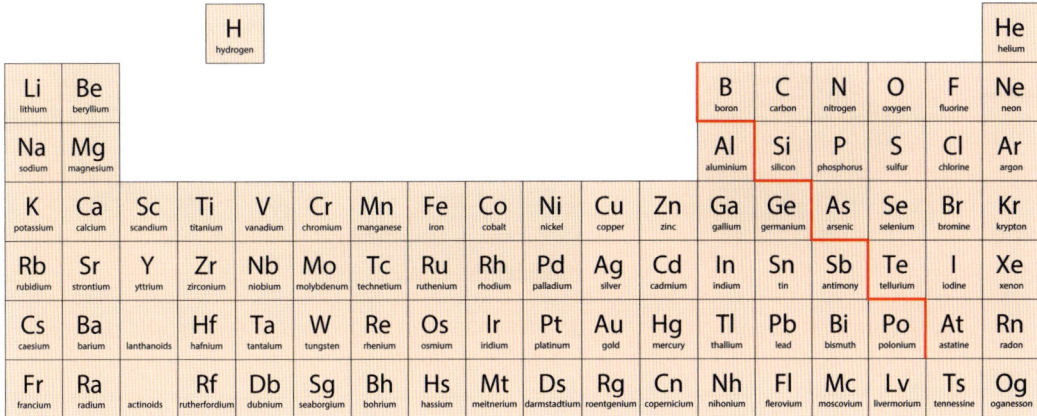

Note: This periodic table does not include all the elements.

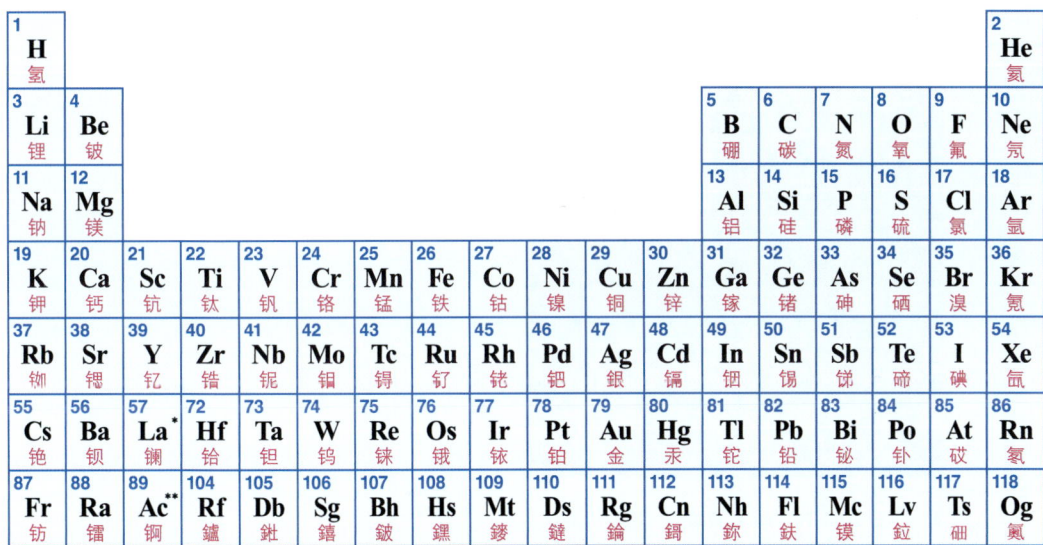

Note: This periodic table does not include all the elements.

▲ This is a Chinese periodic table. It includes international chemical symbols and Chinese element names.

Questions

1. Write the chemical symbols for these elements: carbon, cobalt, nitrogen, neon, magnesium, manganese, and chlorine.

2. Write the names of the elements with these chemical symbols: Ca, Ni, Cr.

3. Find these elements in the periodic table, and write their chemical symbols: hydrogen, helium, lithium, boron, oxygen, fluorine, sodium.

4. Find these chemical symbols in the periodic table, and write their names: V, W, Xe, Y, Zn.

Key points

- The chemical symbol for an element is its one- or two-letter code.
- Scientists all over the world use the same chemical symbols.

41

2.4 Atoms

Objectives

- Define the terms *atom* and *element*
- Describe the strengths and limitations of a model for atoms
- Explain how the properties of an element are the properties of many atoms of the element

Silicon is an element. It holds the tiny electric circuits in every computer, TV, and phone.

▲ A lump of silicon

▲ The integrated circuits in electronic devices are mainly silicon.

The picture below also shows silicon. It was taken with a special microscope. The picture shows the surface of a piece of silicon, magnified ten million times.

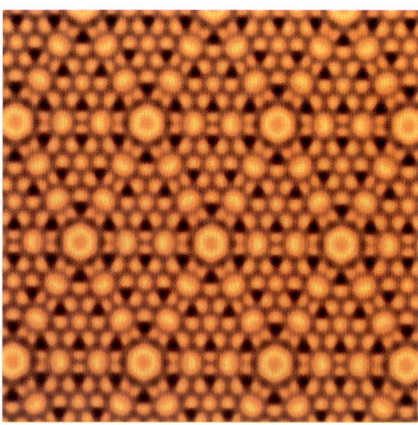

▲ The surface of a piece of silicon, magnified ten million times.

What are atoms?

The spheres in the picture are atoms. Atoms are particles. An **atom** is the smallest particle of an element that can exist.

Atoms are tiny. The diameter of one atom is about 0.000 000 01 cm. If you could place one hundred million atoms side by side, they would stretch one centimetre.

Every element has its own type of atom. For example, silicon atoms are all alike. All platinum atoms are alike. But the atoms of each element are different to the atoms of all other elements. Silicon atoms are different to platinum atoms, carbon atoms, and gold atoms. There are about 100 elements, so there are about 100 types of atom.

You can now give a complete definition of an element. An element is a substance that is made from one type of atom. It cannot be split into other substances.

Elements, compounds, and mixtures

Thinking and working scientifically

How can we model atoms?

Toy bricks can help us imagine atoms.

▲ If these bricks represent gold atoms…

▲ …then this shows a piece of gold.

▲ If these bricks represent oxygen atoms…

▲ …then these represent atoms of another element such as nitrogen.

As you know, in science, a model is an idea that explains observations and helps in making predictions. The bricks help us to imagine the model.

The model has two strengths. It shows that every element has its own type of atom. It also shows that many atoms may join together.

Like all models, the toy brick model is not perfect. It has limitations. For example, atoms do not have straight edges. And atoms are much, much, much smaller than toy bricks.

Does one atom have the properties of an element?

The picture shows a block of copper. Its colour is red-brown. But a single copper atom is not red-brown. A single atom is not shiny. A single atom is not in the solid, liquid, or gas state.

The properties of an element are the properties of many atoms. The copper block weighs 1 kg. It is made up of about 9 000 000 000 000 000 000 000 000 atoms. Together, these atoms make copper red–brown and shiny.

In the copper block, the atoms have fixed positions. If you heat the block to 1084 °C, the atoms leave their positions. The atoms move around, sliding over each other. The copper is melting. One atom of copper cannot melt. Only many atoms together can melt.

▲ A block of copper.

Questions

1. Write the definitions for *atom*, *element*, and *model*.
2. State two properties of copper. Does a single copper atom have these properties?
3. The toy brick model shown above helps you to imagine atoms.
   a. Describe how the toy brick model is useful.
   b. Describe how the bricks in the model are different to real atoms.

Key points

- An atom is the smallest particle of an element that can exist.
- An element is a substance that is made from one type of atom.
- The properties of an element are the properties of many atoms together.

Science in context 2.5

Objectives

- Describe how scientists contributed to the discovery of the periodic table
- Describe how science knowledge may be developed over time

Organising the elements

By 1860 scientists had discovered around 60 elements. They knew about some of their properties. But much of the knowledge was a jumble of facts. They wanted to find out more about them. So they asked scientific questions:

- What are the patterns in the properties of elements?
- How many more elements are there?
- Can we use patterns in properties to help find new elements?

▲ Some of the 60 elements that had been discovered by 1860: iodine, copper, and sulfur.

Building up knowledge

In the early 1800s, John Dalton studied evidence from experiments. He suggested that atoms of each element have a different mass. An Italian scientist, Stanislao Cannizzaro, worked out the masses.

In 1860, Cannizzaro spoke at a conference in Germany. He gave out data about the masses of the atoms of the elements known at the time. A Russian scientist, Dmitri Mendeleev, picked up a copy.

Several scientists tried using the mass data to show patterns in element properties. No one worked out the perfect arrangement.

One day in 1869, Mendeleev was at home, writing a textbook. He was trying to organise information about elements' properties. Mendeleev cut out some cards. On each card he wrote the name of an element and information about the element, including:

- its atomic mass
- its properties.

Mendeleev tried sorting the cards in different ways. Eventually, he came up with an arrangement that worked. The elements were in atomic mass order. Elements with similar properties were grouped together. Mendeleev wrote his arrangement on the back of an envelope. This was the first periodic table.

Mendeleev was confident in his arrangement. He left gaps for elements that he predicted should exist but that had not been discovered. He predicted the properties of these elements.

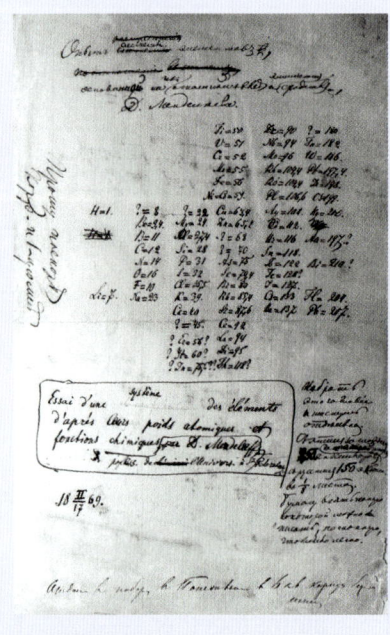

▲ Mendeleev's first periodic table.

44

Elements, compounds, and mixtures

Finding the missing elements

Over the years, scientists tried to find the elements that Mendeleev had predicted.

In 1874, a French scientist found the missing element beneath aluminium. He named it gallium. The properties of gallium were as Mendeleev predicted.

Swedish scientists soon discovered two more of Mendeleev's missing elements – scandium and germanium. Their properties were as Mendeleev predicted.

By 1925, most of the gaps in the periodic table had been filled. But there were spaces beneath manganese for two more elements. Two German scientists, Ida Tacke and Walter Noddack, looked for the missing elements. After months of hard work, they extracted 1 g of a new substance from 660 kg of a rock. It was one of the missing elements. They called it rhenium.

The discoveries of the missing elements showed that Mendeleev's predictions were correct. They gave scientists confidence in Mendeleev's great work.

▲ Mendeleev predicted the existence of the element scandium, and its properties.

▲ Ida Tacke was one of the discoverers of rhenium.

📖 Key points

- Mendeleev used data from other scientists to help develop the periodic table.
- Other scientists found evidence to show that Mendeleev's predictions were correct.

❓ Questions

1. Name three scientists whose work contributed to the discovery of the periodic table.
2. Make a timeline to show how different scientists contributed to the discovery of the periodic table and the missing elements.
3. Suggest why other scientists accepted Mendeleev's periodic table as an explanation of the patterns of the properties of the elements.

2.6 Compounds

Objectives

- Define the terms *compound* and *molecule*
- Describe the difference between elements and compounds
- Explain why elements and their compounds have different properties

The white part of your teeth is enamel. Enamel has atoms of three elements:

- calcium – a shiny element that fizzes in water
- phosphorus – a poisonous element that catches fire easily
- oxygen – an element that is in the gas state at 20 °C.

So why don't your teeth catch fire? Or poison you? Or fizz when you drink water?

▲ Tooth enamel protects your teeth.

▲ Calcium in water

▲ Burning phosphorus

What is a compound?

In tooth enamel, the atoms of calcium, phosphorus, and oxygen are not just mixed up. The atoms are joined together to make one substance, calcium phosphate. This means that calcium phosphate is a compound.

A **compound** is a substance made up of atoms of two or more elements, strongly joined together. The difference between an element and a compound is that an element has one type of atom, but a compound has two or more types of atom.

Why do elements and compounds have different properties?

As you know, the properties of a substance depend on its particles. Elements and compounds have different particles, so their properties are different.

Water

Water is a compound. It is made up of atoms of two elements, hydrogen and oxygen. Hydrogen, oxygen, and water have different properties.

- Hydrogen is in the gas state at 20 °C. It burns easily.
- Oxygen is in the gas state at 20 °C. It helps other substances to burn.
- Water is liquid at 20 °C. It puts out fires.

Water, hydrogen, and oxygen have different properties because their particles are different.

- In hydrogen gas, the atoms join together in pairs. The two-atom particles are molecules of hydrogen. A **molecule** is a particle made up of two or more atoms strongly joined together.
- Oxygen gas also exists as molecules. Each molecule has two oxygen atoms.
- Water particles are three-atom molecules. Each molecule has one oxygen atom joined to two hydrogen atoms.

Elements, compounds, and mixtures

🧪 Thinking and working scientifically

Modelling molecules

The diagrams below model hydrogen, oxygen, and water molecules.

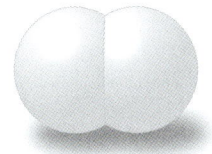

▲ A hydrogen molecule consists of two hydrogen atoms.

▲ An oxygen molecule consists of two oxygen atoms.

▲ A water molecule has one oxygen atom joined to two hydrogen atoms.

The model has strengths and limitations. One strength is that it shows the number and type of atom in each particle. One limitation is that it does not show that hydrogen and oxygen atoms are different sizes.

Carbon monoxide and carbon dioxide

Carbon is an element. It is in the solid state at 20 °C. The element oxygen is vital to life. Carbon and oxygen atoms join together to make two different compounds. The properties of the compounds are different to the properties of the elements, and different to each other.

- A carbon monoxide molecule has one carbon atom joined to one oxygen atom. Carbon monoxide gas is poisonous.

- A carbon dioxide molecule has one carbon atom joined to two oxygen atoms. Carbon dioxide gas is not poisonous. But in the air, it makes the Earth's surface hotter. It is a greenhouse gas.

The two compounds have different properties because their particles are different.

▲ A carbon monoxide molecule

▲ A carbon dioxide molecule

❓ Questions

1. Write the definitions for *compound* and *molecule*.
2. Explain how elements and compounds are different, in terms of their atoms.
TWS 3. The diagrams show particles of elements and compounds.

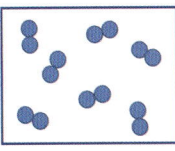 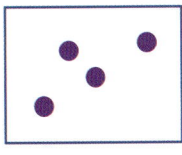 (C - purple particles)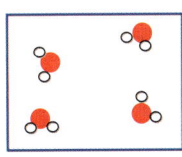

A   B   C   D

   a. Give the letter that shows molecules of an element.
TWS   b. Give the letters that show compounds.
4. Explain why carbon monoxide and carbon dioxide have different properties.

📖 Key points

- A compound is a substance made up of atoms of two or more elements, strongly joined together.
- A molecule is a particle made up of two or more atoms strongly joined together.
- An element has one type of atom, but a compound has two or more types of atom.

47

2.7 What's in a name?

The pictures show two compounds – copper sulfide and copper sulfate.

Objectives

- Write compound names
- Deduce the elements in a compound from the compound name

▲ *Copper sulfide*

▲ *Copper sulfate*

The two compounds have different properties. Their properties are different because their particles are different. The compound names show which elements' atoms are in them.

How do we name compounds?

Compounds of a metal and a non-metal

Copper sulfide has atoms of two elements – copper and sulfur. Copper is a metal and sulfur is a non-metal. Follow these rules to name a compound of a metal and a non-metal.

1. Write the name of the metal element.
2. Write the name of the non-metal element, but change the end of the name to *-ide*.

Metal	Non-metal	Name of compound
copper	sulfur	copper sulfide
sodium	chlorine	sodium chloride
calcium	oxygen	calcium oxide

As you know, the periodic table shows which elements are metals and which are non-metals. Elements to the left of the stepped line are metals. Elements to the right of the stepped line are non-metals.

▲ *Most people call sodium chloride, salt.*

Compounds of a metal, a non-metal, and oxygen

Copper sulfate has atoms of three elements – copper, sulfur, and oxygen. Follow these rules to name a compound of a metal, a non-metal, and oxygen.

1. Write the name of the metal element.
2. Write the name of the non-metal element, but change the end of its name to *-ate*. The *-ate* shows that the compound includes oxygen atoms.

48

Elements, compounds, and mixtures

Elements in compound	Name of compound
copper, sulfur, oxygen	copper sulfate
calcium, phosphorus, oxygen	calcium phosphate
cobalt, nitrogen, oxygen	cobalt nitrate
calcium, carbon, oxygen	calcium carbonate

Compounds of non-metals

Some compounds are made up of atoms of non-metals only. To name these atoms, you need to know the numbers of atoms of each element in a molecule of the compound.

▲ Cobalt nitrate has atoms of three elements – cobalt, nitrogen, and oxygen.

  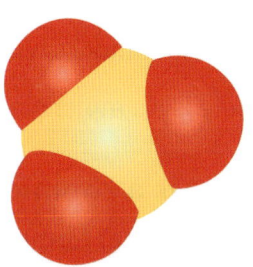

▲ This molecule has one carbon atom and one oxygen atom. The compound name is carbon mon*oxide*.

▲ This molecule has one sulfur atom and two oxygen atoms. The compound name is sulfur di*oxide*.

▲ This molecule has one sulfur atom and three oxygen atoms. The compound name is sulfur tri*oxide*.

▲ The cliffs are made of calcium carbonate. Calcium carbonate has atoms of calcium, carbon, and oxygen.

Questions

1. Write the names of the compounds that are made up of atoms of:
   a. copper and sulfur
   b. sodium and chlorine
   c. copper, sulfur, and oxygen

2. Write the names of the elements in these compounds.
   a. magnesium oxide
   b. potassium chloride
   c. potassium sulfate

3. Predict the names of the compounds that are made up of atoms of:
   a. calcium and chlorine
   b. copper and oxygen
   c. calcium, nitrogen, and oxygen

4. Write the names of the compounds with molecules that are made up of:
   a. One carbon atom and one oxygen atom
   b. One sulfur atom and two oxygen atoms
   c. One nitrogen atom and two oxygen atoms

Key points

- A compound of a metal and a non-metal element has a two-word name. The second word is the name of the non-metal, with *-ide* at the end.

- If a compound name ends in *-ate*, the compound includes oxygen atoms.

- *mon-*, *di-*, or *tri-* means that a molecule contains 1, 2, or 3 atoms of an element.

Thinking and working scientifically

2.8

Objectives
- Write chemical formulae for elements and compounds
- Make deductions from formulae

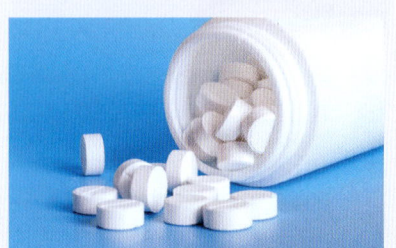

▲ Paracetamol is a painkiller.

▲ TNT is an explosive.

Chemical formulae

Paracetamol is a painkiller. Its molecules are made up of atoms of four elements – carbon, hydrogen, nitrogen, and oxygen. TNT is an explosive. Its molecules are made up of atoms of the same four elements. Why do paracetamol and TNT have different properties?

The two substances have different properties because their particles are different. Their molecules have different numbers of atoms of the four elements:

- A paracetamol molecule has 8 carbon atoms, 9 hydrogen atoms, 1 nitrogen atom, and 2 oxygen atoms.
- A TNT molecule has 7 carbon atoms, 5 hydrogen atoms, 3 nitrogen atoms, and 6 oxygen atoms.

There is a simpler way of showing this information, in chemical formulae.

What is a chemical formula?

The **chemical formula** of a substance gives the relative number of atoms of each element in the substance. It includes the chemical symbols of the elements that are in it. For example:

- The formula of carbon dioxide is CO_2. This shows that there are two oxygen atoms for every one carbon atom.
- The formula of carbon monoxide is CO. This shows that there is one oxygen atom for every one carbon atom.
- The formula of water is H_2O. This shows that there are two hydrogen atoms for every one oxygen atom.

The chemical formulae above are for compounds that exist as molecules. Elements that exist as molecules also have chemical formulae. For example:

- An oxygen molecule is made up of two atoms. Its formula is O_2.
- A phosphorus molecule is made up of four atoms. Its formula is P_4.

   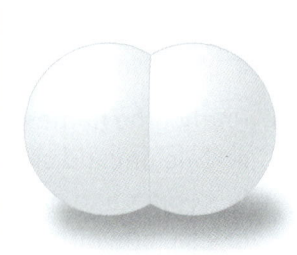

▲ Carbon dioxide, CO_2 ▲ Carbon monoxide, CO ▲ Water, H_2O ▲ Oxygen, O_2

For any substance that exists as molecules, the chemical formula gives the number of atoms of each element in the molecule.

More chemical formulae

Many compounds do not exist as molecules. Here are some examples of their formulae.

Name of compound	Relative number of atoms of each element	Formula
sodium chloride (salt)	1 sodium : 1 chlorine	NaCl
copper sulfide	1 copper : 1 sulfur	CuS
copper sulfate	1 copper : 1 sulfur : 4 oxygen	$CuSO_4$
sodium carbonate	2 sodium : 1 carbon : 3 oxygen	Na_2CO_3

Writing chemical formulae

If you know the relative number of atoms of an element in a substance, you can write its chemical formula. Follow these rules for writing numbers in formulae:

- Write each number to the right of its chemical symbol.
- Write the numbers smaller than the chemical symbols.

For example:

- A sulfur trioxide molecule has one sulfur atom and three oxygen atoms. Its formula is SO_3.
- The relative numbers of atoms in calcium carbonate are: 1 calcium : 1 carbon : 3 oxygen. Its formula is $CaCO_3$.

You can use the information about paracetamol and TNT to write their chemical formulae:

- The formula of paracetamol is $C_8H_9NO_2$.
- The formula of TNT is $C_7H_5N_3O_6$.

Making deductions from chemical formulae

If you know the formula of a substance, you can work out the relative number of atoms of each element in the substance. For example:

- A substance has the formula Br_2. It is an element, bromine. Bromine exists as two-atom molecules.
- A substance has the formula SiO_2. It is a compound. It has two atoms of oxygen to one atom of silicon.

Questions

1. Write the chemical formula of **a** carbon monoxide **b** carbon dioxide.
2. Predict the chemical formula of **a** nitrogen monoxide **b** sulfur dioxide.
3. Choose the chemical formulae that show compounds:
   P_4 P_2O_5 N_2 $CaSO_4$
4. A substance in asthma medicine, salbutamol, exists as molecules. Its chemical formula is $C_{13}H_{21}NO_3$. Give the number of atoms of each element in a salbutamol molecule.

Key points

- The chemical formula of a substance gives the relative number of atoms of each element in the substance.

Elements, compounds, and mixtures

2.9 What's in a mixture?

Objectives

- Define the term *mixture*
- Interpret particle diagrams of mixtures
- Describe some applications of science

Three friends are having a picnic on the beach. Prita spills water in the salt. Prema finds a stone in her rice. Priyam crunches sand in her salad.

What are mixtures?

Salt and water, stones and rice, and sand and salad are mixtures. A **mixture** contains two or more substances that are not joined to each other. They are just mixed up.

Often, it is easy to separate the substances in a mixture. Prema could pick stones out of the rice. Priyam could wash sand off the salad. Prita could evaporate water from the salt.

> ### Science in context
>
> **Useful mixtures**
>
> Many useful things are mixtures. Scientists mix together exactly the right amounts of substances to make paints, medicines, and deodorants. Alloys are mixtures that are mainly metals. You can learn about alloys in Chapter 3.

Inside mixtures

Mixtures can contain elements, compounds, or both.

- The jar contains iron and copper nails. Iron and copper are elements, so the jar contains a mixture of elements.
- The ingredients in toothpaste are compounds, so toothpaste is a mixture of compounds.
- The air is a mixture of elements (including nitrogen, oxygen, and argon) and compounds (including carbon dioxide and water).

▲ Paints, medicines, and deodorant are mixtures.

▲ A mixture of elements

◀ Toothpaste is a mixture of compounds.

52

Elements, compounds, and mixtures

Thinking and working scientifically

Modelling mixtures

You can use particle diagrams to model mixtures. In the diagrams, each circle is one atom. Atoms of different elements are shown in different colours.

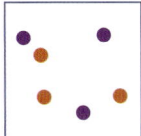
▲ This is a mixture of two elements in the gas state. Both elements exist as single atoms.

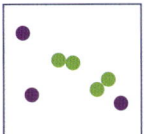
▲ This is a mixture of two elements. One element exists as single atoms. The other element exists as molecules.

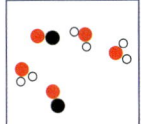
▲ This is a mixture of two compounds.

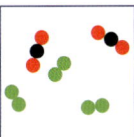
▲ This is a mixture of an element and a compound.

This type of model has strengths and limitations. Its strength is that it clearly shows the atoms in the elements and compounds. Its limitation is that it is not to scale.

What are mixtures like?

In any mixture:

- The different substances are not joined together.
- The substances keep their own properties.
- You can change the amounts of the substances.
- It is often easy to separate the substances.

Questions

1. Write the definition for *mixture*.

2. Name four useful mixtures.

3. The diagrams show particles in mixtures.

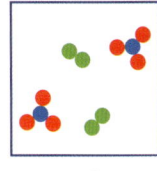 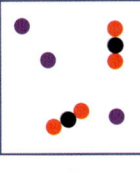 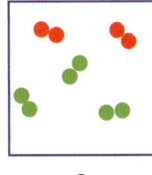 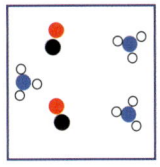
    A   B   C   D

  a. Give the letter that shows a mixture of two elements.

  b. Give the letter that shows a mixture of two compounds.

  c. Give the **two** letters that show mixtures of an element and a compound.

4. A mixture includes substances with these formulae: CO_2, N_2, SO_3, O_2.

  a. Give the number of substances in the mixture.

  b. Write the names and formulae of the elements in the mixture.

  c. Write the names and formulae of the compounds in the mixture.

Key points

- A mixture contains two or more substances that are not joined together.
- The substances in a mixture keep their own properties.

2.10 What's in a solution?

Kasarna makes coffee. She pours hot water over coffee powder. Then she adds sugar and stirs. The sugar and coffee powder dissolve in the water. Kasarna has made a solution.

Objectives

- Define the terms *solution*, *soluble*, and *insoluble*
- Explain why a solution is a mixture
- Evaluate a model for a solution
- Make a prediction and conclusion

What's in a solution?

A **solution** is a mixture that forms when a substance dissolves in a liquid. All parts of a solution are the same. In a solution:

- the liquid is the **solvent**
- the substance that dissolves is the **solute**.

In a solution, particles of the solute and solvent are mixed randomly. They move around, sliding over each other.

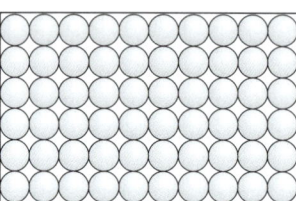
▲ Particles in solid sugar.

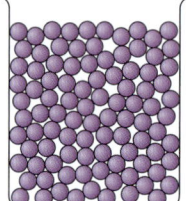
▲ Particles in liquid water.

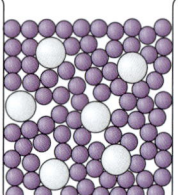
▲ Particles in sugar solution.

Not to scale. In these diagrams, each circle represents one particle. The particles are molecules.

- A water molecule has one oxygen atom joined to two hydrogen atoms.
- A sugar molecule has 12 carbon atoms, 22 hydrogen atoms, and 11 oxygen atoms.

▲ Making coffee

🧪 Thinking and working scientifically

Modelling solutions

You can use rice and beans to model particles in a solution. In the image, rice grains represent water particles. Beans represent sugar particles.

Can you think of some strengths and limitations of the rice and beans model?

▲ Rice and beans can model particles in a solution.

Why is a solution a mixture?

Amun adds sugar to water. He stirs the mixture. The sugar dissolves to make a solution. The solution is a mixture because:

- It contains particles of more than one substance that are not joined together.
- The amounts of substances in the solution can change.
- It is easy to separate the water and sugar, by letting the water evaporate.

▲ Adding sugar to water.

- The substances in the solution keep their own properties. Amun can taste the sugar in the solution, for example.

Thinking and working scientifically

What happens to the sugar?

When Amun adds sugar to water, the sugar seems to disappear. Amun wants to check that the sugar is still there, without tasting the solution.

Planning the investigation

Amun plans an investigation. He decides to:

- Find the mass of a beaker of water.
- Weigh out 10 g of sugar.
- Add the sugar to the water in the beaker, and stir. The sugar will dissolve.
- Find the mass of the beaker and its solution.

▲ *The mass of the solution on the left is the same as the total mass of sugar and water on the right.*

Amun makes a prediction, based on science knowledge:

The total mass of the beaker, water, and sugar at the start will be equal to the mass of the beaker and sugar solution at the end. I think this because there are the same number of water and sugar particles before and after mixing.

Carrying out the investigation

Amun follows his plan. He writes the results in a table.

What I weighed	Mass (g)
water + beaker (at start)	300
sugar	10
water + sugar solution (at end)	310

Analysing the evidence

Amun makes a conclusion:

My prediction was correct. The total mass of the beaker of water and sugar at the start (300 g + 10 g = 310 g) is equal to the mass of the beaker and sugar solution at the end (310 g). My results show that all the sugar dissolves in the solution. The sugar particles have not disappeared.

Amun could improve his conclusion by describing a limitation. For example, he only investigated one solute, sugar. He does not know if the same conclusion would be true for other solutes, like salt.

Do all substances dissolve?

Sugar and salt dissolve in water. They are **soluble** in water. Some substances, such as sand and chalk, do not dissolve in water. They are **insoluble** in water.

Questions

1. Write the definitions for *solution*, *soluble*, and *insoluble*.
2. Name the solute and solvent in a solution of sugar in water.
3. Explain why a solution is a mixture. Give four reasons in your answer.

Key points

- A solution is a mixture made when a substance dissolves in a liquid.
- Solutions are mixtures because they contain particles of more than one substance.

2.11 Comparing elements, mixtures, and compounds

Objective

- Describe differences between elements, mixtures, and compounds

Sodium is a shiny metal. It fizzes in water – and may even explode. Chlorine is a green, smelly, poisonous gas at 20 °C. A mixture of sodium and chlorine can kill. But when the two elements join together, they make a compound. The compound is sodium chloride, the salt you may add to food.

▲ Sodium may explode in water.

▲ Chlorine gas is poisonous.

▲ You may add sodium chloride (salt) to food.

Elements, a mixture, and a compound

Elements

Ricardo collects 7 g of iron powder and 4 g of sulfur powder. Iron is an element. It has one type of atom. It cannot be split into other substances. Ricardo holds a magnet near the iron. The iron moves onto the magnet.

Sulfur is also an element. It has one type of atom, and cannot be split into other substances. Its atoms are different from iron atoms.

▲ Iron (an element)

▲ Sulfur (an element)

A mixture

Ricardo mixes the iron and sulfur. He can see the two elements. He can separate them with a magnet. The properties of the mixture are similar to the properties of the elements. The elements have not joined together, so it is easy to separate them. There can be any amounts of iron and sulfur in the mixture.

▲ A magnet separates a mixture of iron and sulfur.

Elements, compounds, and mixtures

A compound

Ricardo heats the mixture. The two elements join together to form a compound, iron sulfide. Ricardo cannot get iron out of the compound. The properties of the compound are different from the properties of the elements.

The amounts of the elements in a compound are important. Iron sulfide, FeS, always has 7 g of iron for every 4 g of sulfur.

▲ *Iron sulfide (a compound)*

How are elements, mixtures, and compounds similar and different?

The table summarises the differences between elements, mixtures, and compounds.

	Element	Mixture	Compound
Is it a single substance?	yes	no	yes
How many types of atom are in it?	one	two or more	two or more
Can it be split up or separated?	no	yes	not easily
What are its properties?	an element has its own properties	the substances in the mixture keep their own properties	a compound has different properties to the properties of the elements in it
How much of each substance is in it?	—	you can change the amounts of substances	the relative amounts of each element cannot change
Are its substances joined together?	—	no	yes, atoms of the elements are joined together

❓ Questions

1. Describe two ways that elements and compounds are similar.
2. Describe two ways that elements and compounds are different.
3. Describe two ways that compounds and mixtures are different.
4. The diagrams show particles in elements, compounds, and mixtures.
   a. Give the letter that shows a compound.
   b. Give the letter that shows an element.
   c. Give the letter that shows a mixture of two elements.
   d. Give the letter that shows a mixture of an element and a compound.

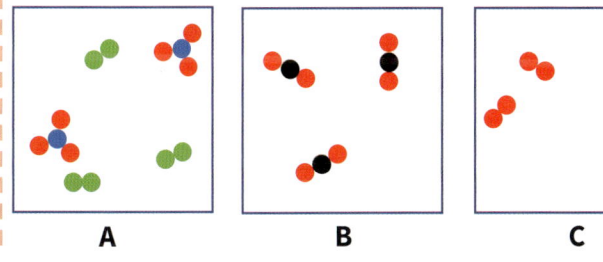

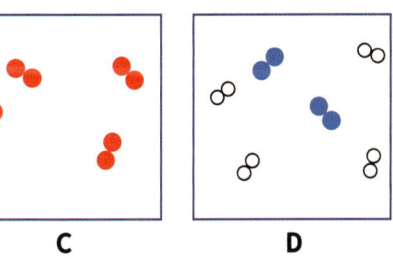

A   B   C   D

 **Key points**

- Elements, mixtures, and compounds are different.

57

Extension 2.12

Objective

- Name the main elements in living things

▲ Asim has a mass of 50 kg.

What are you made of?

Inside Asim

Asim has a mass of 50 kg. His body contains:

- enough hydrogen to fill his classroom
- enough oxygen to fill a room in his house
- enough nitrogen to fill his school bag 85 times
- and enough carbon to make a huge number of pencils.

Asim's body also contains 0.5 kg of phosphorus and small amounts of many other elements. The elements in Asim's body are not just mixed up. They are joined together in hundreds of different compounds.

Asim's blood is a mixture of compounds, including water. As you know, water is a compound of hydrogen and oxygen.

Asim's nails are mainly keratin. Keratin is a compound. It is made of atoms of carbon, hydrogen, oxygen, and nitrogen. There are also atoms of sulfur, which makes keratin hard and rigid.

All Asim's body tissues and organs – including his skin, bones, brain, and heart – are compounds made mainly of carbon, hydrogen, oxygen, and nitrogen.

Body substances – where do they come from?

Everyone needs the element oxygen. The air you breathe contains 21% oxygen mixed with other gases. Your lungs separate oxygen from these gases.

Everyone needs water, too. It comes from our food and drink.

As well as water and oxygen, you need proteins, fats, and carbohydrates. These are all compounds. They are in your food.

▲ Food contains proteins, fats, and carbohydrates.

Other important chemicals

To keep healthy, your body needs small amounts of **vitamins**. Vitamins are compounds made up mainly of the elements carbon, hydrogen, and oxygen.

You also need small amounts of other elements, such as iron and calcium. But it's no use swallowing iron nails or lumps of calcium metal – and it would be dangerous to do so. You need compounds that contain these elements, called **minerals**.

Mineral deficiency

If you do not take in enough of any mineral, you may suffer symptoms of mineral deficiency.

Mineral	Symptoms of deficiency
iron	tiredness, lack of energy, shortness of breath
calcium	weak bones and frequent fractures
zinc	reduced growth in children, problems with senses and memory
iodine	swelling of thyroid gland in neck, tiredness, brain damage

Science in context

Iodine deficiency

In the 1980s, about 25% of people in Tanzania had iodine deficiency disorders. The government told salt makers to add iodine to all salt sold in Tanzania.

The government wanted to know if its policy had worked. In 2004, scientists investigated two questions:

- What percentage of households use salt with added iodine?
- Did fewer people suffer from iodine deficiency in 2004 than in the 1980s?

The scientists gathered evidence. They tested salt samples from 156 000 households. Iodine had been added to more than 80% of the salt samples.

The scientists tested 166 000 children for iodine deficiency symptoms. The percentage of children with iodine deficiency had decreased from 25% in the 1980s to 7% in 2004.

The scientists studied their evidence and made conclusions. In most areas, the greater the number of people eating iodised salt, the smaller the number of people with iodine deficiency.

The scientists advised the government to make sure that iodine is added to all salt in future.

▲ This woman has not had enough iodine in her diet.

Questions

1. Name the four elements that make up most of the mass of your body.
2. Describe what may happen if you do not take in enough iron.
3. Give the symptoms of calcium deficiency, iodine deficiency, and zinc deficiency.

Key points

- Your body is made up of compounds containing mainly the elements carbon, hydrogen, and oxygen.
- A healthy diet includes proteins, carbohydrates, fats, vitamins, and minerals.

Review 2.13

1. Choose one word from the list below to answer each question part.

solvent	molecule
compound	solution
element	atom
model	chemical symbol

   a. A substance that is made from one type of atom and that cannot be split into other substances. [1]

   b. A way of representing something that you cannot see or experience directly. [1]

   c. The smallest particle of an element that can exist. [1]

   d. The one- or two-letter code for an element. [1]

   e. A substance made up of atoms of two or more elements, strongly joined together. [1]

   f. A particle made up of two or more atoms, strongly joined together. [1]

   g. A mixture that forms when a substance dissolves in a liquid. [1]

   h. The liquid that a solute dissolves in. [1]

2. Copy and complete the table. Use the periodic table to help you. [8]

Element name	Chemical symbol
	Be
	B
chlorine	
fluorine	
potassium	
silicon	
sodium	
sulfur	

3. Each of these chemical symbols has one mistake. Write the chemical symbols correctly.

   a. HE [1]
   b. AR [1]
   c. ne [1]
   d. b [1]

4. Copy and complete the sentences below. Use these phrases. You can use each phrase once, more than once, or not at all.

   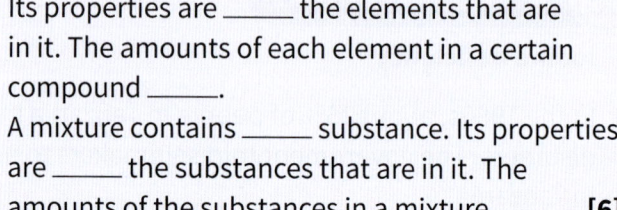

one	similar to
more than one	can vary
different from	are always the same

   A compound is made up of _____ type of atom. Its properties are _____ the elements that are in it. The amounts of each element in a certain compound _____.
   A mixture contains _____ substance. Its properties are _____ the substances that are in it. The amounts of the substances in a mixture _____. [6]

5. Jav is using toy bricks to model elements and compounds.

   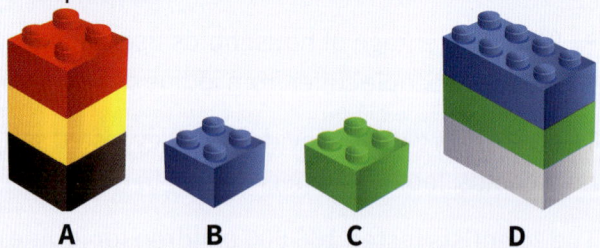

   A   B   C   D

   a. Write **two** letters that model elements. [1]

   b. Write **two** letters that model compounds. [1]

   c. Give **two** ways in which **D** is a useful model for the type of particle it represents. [2]

   d. Give one way in which **D** is not like the particle it represents. [1]

6. Copy and complete the table. [6]

Compound	Names of elements in compound
calcium oxide	
sodium chloride	
magnesium sulfide	
iron sulfate	
calcium carbonate	
silver nitrate	

7. Write the names of the compounds with the formulae below.

a. NO_2 [1]

b. NO [1]

c. SO_3 [1]

d. $MgSO_4$ [1]

e. $NaCl$ [1]

f. $CaCO_3$ [1]

g. BeO [1]

8. The picture shows a model of a molecule.

a. Give the total number of atoms in the molecule. [1]

b. Give the number of different elements whose atoms are in the molecule. [1]

c. The model is an oxygen difluoride molecule.

   i. Deduce the colour of the fluorine atoms in the model. [1]

   ii. Deduce the chemical formula of oxygen difluoride. [2]

9. Copy and complete the table to show whether each substance is an element or a compound. [6]

Name of substance	Formula	Element or compound?
nitrogen	N_2	
carbon dioxide	CO_2	
sulfur	S_8	
argon	Ar	
magnesium oxide	MgO	
copper sulfate	$CuSO_4$	

10. The picture shows a crystal of citrine. Citrine is mainly silicon dioxide.

a. Give the formula of silicon dioxide. [1]

b. Explain whether silicon dioxide is an element or a compound. [2]

11. The diagrams show some particles of single substances and mixtures of substances in the gas state.

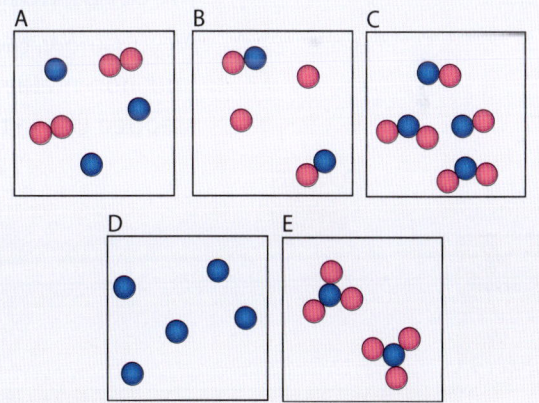

a. Write the letter of the diagram that shows a single element. [1]

b. Write the letter of the diagram that shows a single compound. [1]

c. Write the letter of the diagram that shows a mixture of elements. [1]

d. Write the letter of the diagram that shows a mixture of compounds. [1]

e. Write the letter of the diagram that shows a mixture of an element and a compound. [1]

3.1 Magnificent metals

Objectives

- Give six physical properties of typical metal elements
- Explain how the uses of metal elements depend on their properties

Would you like to watch football in this aluminium-covered stadium? Or visit the copper-clad Museum of Fire? The properties of aluminium and copper are perfect for these buildings.

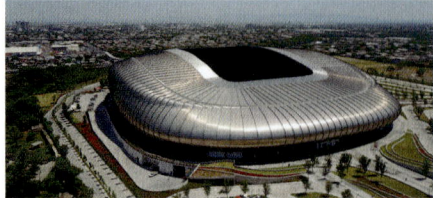

▲ The aluminium-clad Monterrey stadium, Mexico.

▲ The copper-clad Museum of Fire, Poland.

What are properties?

As you know, every material has its own **properties**. The properties of a material describe what it is like and what it does. Aluminium is shiny, for example, and copper is orange coloured.

There are two types of properties:

- As you know, **physical properties** are properties that you can observe or measure without changing the material.
- **Chemical properties** describe how substances change in chemical reactions. You will learn about chemical properties in Chapter 4.

Metal or non-metal?

As you know, the chemical elements can be split into two groups: metals and non-metals. Most elements are metals. Metals are on the left of the stepped line in the periodic table. Aluminium and copper are metals.

Note: This periodic table does not include all the elements. ▲ The periodic table. Metals are on the left of the stepped line.

What makes metals useful?

Every element has its own properties. But metals have similar physical properties to each other. Their properties make them useful.

Metals and non-metals

State and appearance

Most metal elements have high melting points. They are solid at room temperature. Only mercury is different. It is liquid at room temperature.

All metal elements are shiny when you first cut them, or if you rub them with sandpaper. After a while, many metals lose their shine. But gold and platinum are always shiny. They make beautiful jewellery.

Sonority

When you hit a metal, it makes a ringing sound. Scientists say that metals are **sonorous**.

Conducting heat and electricity

Thermal energy and electricity travel through metals easily. Metals are good **conductors** of thermal energy and electricity. Some metals are better conductors than others:

- Copper is an excellent conductor of thermal energy. Copper heat sinks stop computers overheating. The copper transfers thermal energy away from the processing unit.
- Aluminium and gold are excellent conductors of electricity. Aluminium is used for power cables. Gold is used in printed circuit boards in cell phones.

▲ A copper heat sink

▲ Aluminium power cables

▲ Gold in a cell phone

Changing shape

Thin metal sheets are bendy. So, when a car crashes, its metal body does not break into little pieces – it just bends. Metals are also:

- **malleable** – they can be hammered into different shapes
- **ductile** – they can be pulled into wires.

Strength

Many metal elements are **strong**. You need big forces to break them. Buses and ships need to be strong. They are mainly iron.

▲ Skilled people made this necklace in Egypt 3500 years ago. It is still shiny.

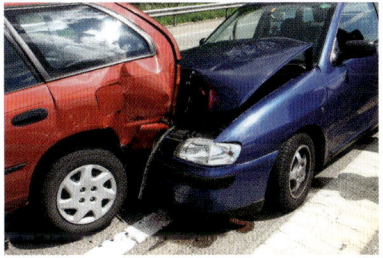
▲ Metals are sonorous, so they make good bells.

▲ Metal bends when cars crash.

Questions

1. Write a definition of *physical properties*.
2. List six physical properties of a typical metal element.
3. Explain why power cables are made of aluminium.
4. Platinum is a typical metal element. Suggest two physical properties of platinum that explain why it makes good jewellery.

📖 Key points

- Most metals are shiny, sonorous, strong, malleable, and ductile.
- Most metals are good conductors of thermal energy and electricity.
- The uses of metals depend on their properties.

Thinking and working scientifically 3.2

Comparing conductors

What are heat sinks?

Tuqa's family has a computer repair business. Tuqa knows that computers have metal heat sinks. A heat sink transfers thermal energy away from the processor. It stops the computer getting too hot.

▲ An aluminium heat sink.

▲ A copper heat sink.

Tuqa notices that some computers have aluminium heat sinks. Other computers have copper heat sinks. She wonders which metals make better heat sinks. She also wonders why she never sees iron or zinc heat sinks.

Planning the enquiry

Asking a scientific question

Tuqa knows that *Which heat sink is better?* is not a scientific question. She makes up a question that can be answered using evidence. Her scientific question is:

Which conducts thermal energy quickest – aluminium, copper, iron, or zinc?

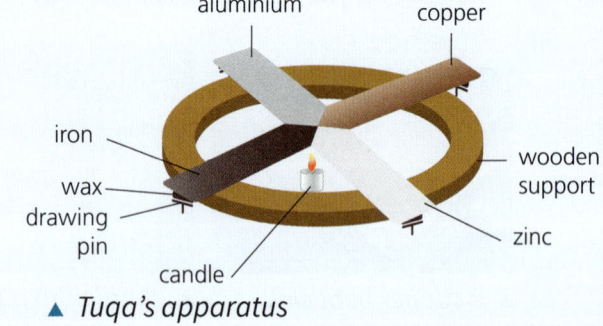

▲ Tuqa's apparatus

Planning the investigation

Tuqa has the apparatus in the diagram.

She plans to light the candle. Thermal energy from the candle will travel along the metal strips. The wax will melt and the pins will drop off. She will measure the times from lighting the candle to the pins falling off.

Tuqa knows that changing one variable will affect another variable. This means that she can answer her question by doing a fair test. She lists the variables:

Objectives

- Define the term *scientific question*
- Describe how to plan a fair test
- Describe how to make accurate and reliable measurements
- Make a conclusion

Type of variable	Meaning	Variable in this investigation
independent variable	the quantity she changes	type of metal
dependent variable	the quantity that changes when the independent variable changes	time for pin to fall off
control variables	variables to keep the same	• size of metal strip • amount of wax • amount of heat

64

Tuqa sets up the apparatus. She makes sure the control variables are the same by:

- measuring the strips to check their size
- weighing out pieces of wax of the same mass.

The same candle heats the four metals at the same time, so the thermal energy supplied is the same for each metal.

Tuqa also plans how to work safely. She will take care when lighting the candle. She will let the hot metals cool before she touches them.

Making a prediction

Before starting, Tuqa wants to make a prediction. She looks in a data book. She finds these data:

Element	Thermal conductivity in W/m/K
aluminium	2.4
copper	3.9
iron	0.8
zinc	1.1

The data show that copper conducts thermal energy quickest. Tuqa predicts that the pin on the copper strip will drop off first.

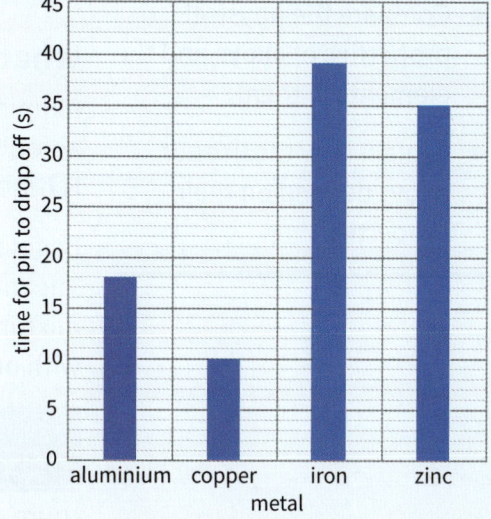

▲ Tuqa's graph.

Carrying out the investigation

Tuqa wants to make accurate measurements. She uses the timer on her phone. She starts the timer when she lights the candle. She notes the time when each pin falls on the desk.

Tuqa wants to her measurements to be reliable. She does the whole experiment three times, and calculates mean values for each reading.

Analysing the evidence

Tuqa carries out her investigation. She analyses her results and plots a bar chart. Then Tuqa uses her bar chart to make a conclusion:

The bar chart shows that the pin drops off the copper first. This shows that copper conducts thermal energy best. My prediction was correct.

The experiment shows why heat sinks are made from aluminium or copper, not iron or zinc.

Questions

1. Write definitions for *scientific question, independent variable, dependent variable, control variable*.
2. Name the type of variable that must not change in a fair test.
3. List the metals in the bar chart in order of how quickly they conduct thermal energy. Start with the slowest.

Key points

- A scientific question is one that is answered using evidence.
- In a fair test, you change the independent variable, measure the dependent variable, and keep the control variables the same.
- Accurate data is data that is close to the true value.
- You can obtain reliable data by making enough measurements.

3.3 Amazing alloys

Objectives

- Define the terms *alloy* and *steel*
- Compare the physical properties of alloys and the elements in them
- Explain how the uses of alloys depend on their properties

Aeroplanes: heavy or light?

Aeroplanes burn huge amounts of fuel. Fuel is expensive. And burning fuel makes greenhouse gases. The lighter an aeroplane, the less fuel it burns. So engineers make aeroplanes from metals that are light for their size.

What is an alloy?

Aluminium is a metal. It is light for its size. This makes it suitable for aeroplanes. But aluminium is not strong, so it breaks easily. And aluminium is not hard, so it is easy to scratch and dent.

Engineers solve these problems by mixing aluminium with other elements. The mixture has different physical properties from pure aluminium. The mixture is stronger and harder than aluminium alone. A mixture of a metal with one or more other elements is an **alloy**.

Material	Composition	Relative hardness	Strength when pulled (MPa)
pure aluminium	100% aluminium	23	8
aluminium alloy 7075	90% aluminium 6% zinc 2% magnesium 1% copper 1% other metals	150	572

▲ Table 1. Properties of aluminium and an aluminium alloy.

▲ The New Aswan Bridge, Egypt, has steel cables.

Steel – a vital alloy

Steel makes many things, from stunning structures to tiny components.

Steels are alloys of iron. Iron is a metal. It is plentiful and cheap. But iron alone is too soft and bendy to be useful. In steels, iron is mixed with small amounts of a non-metal, carbon, and – in some cases – other metals. These elements change the properties of iron and make it more useful.

There are many types of **steel**. Each has its own properties and uses. Table 2 at the top of the next page gives information about three types of steel.

Name of alloy	Other elements	Properties	Uses
low carbon steel	carbon	strong, easily shaped	bridges, buildings, ships, vehicles
manganese steel	manganese carbon	hard tough	mining equipment, railway points
stainless steel	chromium nickel carbon	does not rust	knives and forks, surgical instruments

▲ Table 2. Properties of three types of steel (iron alloys).

Explaining alloy properties

As you have seen, alloys have different physical properties from the elements that are mixed to make them. Alloys are stronger and harder, for example.

The different physical properties of alloys result from their atom arrangements. In a pure metal, the atoms are arranged in layers. The diagram shows part of the atom arrangement in iron. The layers slide over each other easily, so iron is soft and not very strong.

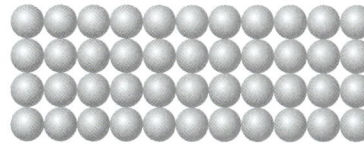
▲ Particles in iron

Carbon atoms are smaller than iron atoms. In steel, carbon atoms get between the iron atoms. The iron atom layers cannot slide over each other easily. This makes steel harder and stronger than pure iron.

An alloy may also have different chemical properties from the elements that are in it. For example, pure iron goes rusty but stainless steel does not. You can read about chemical properties in Chapter 4.

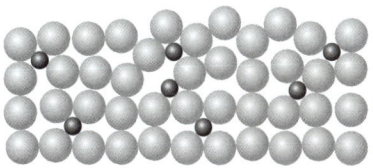
▲ Particles in steel (iron atoms shown grey; carbon atoms shown black)

Bronze – an ancient alloy

Bronze is an alloy of copper and tin. It was first used about 6000 years ago to make tools and weapons. Bronze is harder than stone, which was used before.

▲ An ancient bronze axe

▶ Some motors and gears are made of bronze. They do not need oiling, because bronze has low friction with other metals.

Questions

1. Write definitions of *alloy* and *steel*.
2. Compare the physical properties of pure aluminium and an aluminium alloy. Use the data in Table 1.
3. Explain how the uses of low carbon steel are linked to its physical properties. Use data from Table 2.
4. Explain why pure iron has different physical properties from its alloys. Include a diagram in your answer.

📖 Key points

- An alloy is a mixture of a metal with small amounts of other elements.
- An alloy has different properties from the elements in it.
- Different sized atoms in an alloy change its physical properties.

3.4 Non-metal elements

Objectives

- Define the terms *brittle* and *insulator*
- Give five physical properties of non-metals
- Compare the properties of metals and non-metals
- Deduce whether an element is a metal or a non-metal

Medical oxygen

Oxygen is an important medical treatment. It helps people to breathe.

Metal or non-metal?

Oxygen is a non-metal. A non-metal is an element that is not a metal. Only about 20 elements are non-metals. They are on the right of the stepped line of the periodic table.

In science, only elements can be called non-metals. For example, wood is not a metal. But you cannot say that wood is a non-metal, because wood is not an element.

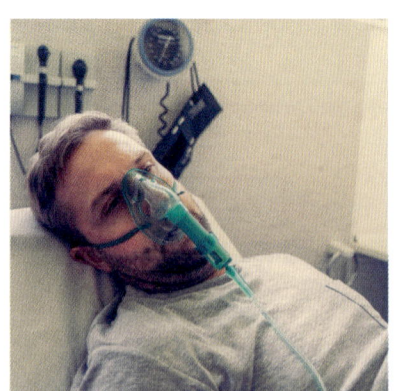

Note: This periodic table does not include all the elements.

▲ The periodic table. Non-metals are on the right of the stepped line.

Non-metal properties

As you know, every element has its own properties. The physical properties of non-metals are similar. They are different from the properties of metals.

▲ These balloons are filled with helium gas.

▲ Neon gas makes this sign glow brightly.

▲ Chlorine gas from the cylinder is bubbled through water. Chlorine makes water safe to drink or swim in.

Non-metals in the gas state

Compared to metals, most non-metals have low melting points and boiling points. Most are in the gas state at room temperature (20 °C).

Non-metals in the solid state

A few non-metals – such as sulfur, carbon, and phosphorus – are in the solid state at 20 °C. They have similar physical properties.

Most non-metals are not shiny. They are dull. Most non-metals are **brittle**, which means that they break easily when hit with a hammer. You cannot bend solid non-metals. Most non-metals do not conduct electricity. They are electrical **insulators**.

Carbon is a special non-metal. There are several types of solid carbon, including graphite and diamond. In each type of carbon, the atoms have a different arrangement. This means that each type of carbon has its own physical properties.

▲ Sulfur is a non-metal. As a solid, it is dull, brittle, and an electrical insulator.

▲ Graphite is soft. It makes the 'lead' in pencils.

Comparing metals and non-metals

The table shows the physical properties of metal and non-metal elements. You can use these properties to work out whether an element is a metal or a non-metal.

Property	Metals	Non-metals
state at 20 °C	most are solid	most are in the gas state
appearance when solid	shiny	dull
do they conduct electricity?	yes	no
other physical properties	strong	brittle
	hard	soft
	malleable and ductile	not malleable or ductile

▲ Properties of metals and non-metals

▲ This is a dentist's drill. It has a diamond tip. Diamond is very hard.

Questions

1. List five physical properties of a typical non-metal element.
2. Explain why scientists do not say that glass is a non-metal.
3. Describe three properties of sulfur that are typical of non-metals, and one property of sulfur that is not typical of non-metals.
4. Element X is solid at 20 °C. It is shiny and conducts electricity. Deduce whether element X is a metal or non-metal.
5. Write a paragraph to compare the physical properties of metal and non-metal elements.

Key points

- Most non-metals have low melting and boiling points.
- In the solid state, most non-metals are brittle and dull.
- Most non-metals do not conduct electricity.

Extension 3.5

Explaining metal and non-metal properties

Objective
- Explain differences in the physical properties of metals and non-metals

▲ Sheets of steel

▲ Sulfur

The train is made from steel, which is an alloy of iron. Iron is a metal element. What properties of the metal make it suitable for trains? Why could you not make a train from a non-metal element, such as sulfur?

Explaining metal and non-metal properties

You can explain the different physical properties of metal and non-metal elements by considering:

- how the particles are arranged
- how strongly the particles are held together.

Melting points

Metals have high melting points. Most are in the solid state at 20 °C. They have this property because their atoms are held together strongly.

Most non-metal elements have low melting points. This is because most exist as single atoms or molecules:

- In a substance that exists as single atoms, the atoms are attracted to each other only weakly.
- For a substance that exists as molecules, the molecules are attracted to one another only weakly.

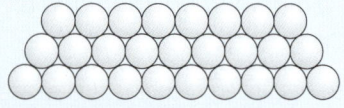

▲ Part of the structure of a typical metal

▲ Helium, neon, argon, and krypton exist as single atoms.

▲ An oxygen molecule is made up of two atoms. Inside a molecule, the atoms are held together tightly. But the molecules are attracted to one another only weakly.

Strength

As you know, metals are strong, especially when they are mixed with other elements in alloys. This is because the atoms in metal elements are held together strongly.

In the solid state, non-metal elements are brittle. Iodine, for example, exists as two-atom molecules. In an iodine crystal, the molecules are arranged in a pattern. But the molecules are attracted to one another only weakly. If an iodine crystal drops to the floor, it breaks between one row of molecules and another. All the broken pieces have straight edges.

▲ Iodine is a non-metal element. Its crystals are brittle.

Key points

- Metals have higher melting points than non-metals because the particles in metals are held together more strongly.
- Metals are stronger than non-metals because the particles in metals are held together more strongly.

Questions

1. Give two factors that affect the melting points and strength of elements.
2. Explain why metals have higher boiling points than non-metals.
3. Explain why most metal elements are strong, but non-metals are brittle.
4. The table shows some properties of two elements, X and Y.

Element	Melting point (°C)	Strength
X	962	strong
Y	−101	weak

   a. Which element might be silver? Explain your decision.
   b. Predict which element conducts electricity better. Explain your decision.

71

Science in context 3.6

Objectives

- Describe some ways that science is applied in society and industry
- Evaluate material properties

Better bicycles

Zaid is buying a new bike. In the shop, he sees bikes made from steel, aluminium, titanium, and carbon fibre. Which should he choose?

Iron or steel?

Many years ago, bikes were made from iron. But the properties of iron are not ideal. The pure metal is soft. It is not very strong.

Bicycle makers tried using different materials, like steel. As you know, steel is an alloy of iron. It is a mixture of iron with carbon and other elements. By 1900, bicycle makers had discovered that steel is strong enough to make into tubes. Soon, bicycles were much less heavy.

Today, many bicycles are made from steel. The alloy is cheap and strong. In a collision, it bends instead of breaking. It is easy to mend steel bikes.

Alternative alloys

The properties of steel are not perfect for bikes. Steel may rust. It is also heavy for its size.

Scientists used data to work out which materials might be better than steel. They made bikes from these materials. They tested the bikes. Today, you can choose between bikes made from steel, or aluminium alloys, or titanium alloys.

▲ Some steels rust if they are not oiled or painted.

The table shows some properties of two alloys. One alloy is mainly titanium. The other alloy is mainly aluminium.

▲ This bicycle has a titanium alloy frame.

Property	Titanium alloy	Aluminium alloy
density, in g/cm^3 (the higher the density, the heavier the bike for its size)	4.4	2.7
relative hardness	334	30
strength, in MN/m^2 (the pulling force that breaks 1 m^2 of the alloy)	950	124
stiffness, in GN/m^2 (the stiffer the alloy, the less it bends)	114	69
fatigue strength, in MN/m^2 (the smallest force that can damage the alloy)	240	62
melting temperature, in °C (the lower the melting temperature, the cheaper the metal is to melt when making the frame)	Up to 1660	Up to 652

Both alloys have advantages and disadvantages. The titanium alloy is stronger, but more expensive. Bikes made from the aluminium alloy are light, but more easily damaged.

Another new material: CFRP

Carbon is a typical non-metal. It is brittle. It breaks if you try to bend it. In the 1950s, scientists tried making carbon threads. They placed them in a resin. They had made a new material: carbon fibre reinforced polymer (CFRP).

Bicycle makers investigated the properties of CFRP. Might it be suitable for bikes? They made some bikes to test their idea. The early CFRP bikes were not perfect, but the scientists persevered. Today, most professional cyclists ride CFRP bikes. They are lightweight and smooth to ride.

Brilliant bamboo

Scientists never stop searching for better bike materials. A recent innovation uses a well-known material: bamboo. A bamboo bike is durable and strong. It is comfortable to ride. Best of all, it is made from a sustainable material.

▲ A bamboo bicycle

▲ Bamboo is a sustainable material.

Questions

1. List three advantages of bamboo bikes.
2. Describe some steps scientists may use to discover new bicycle materials.
3. Write a paragraph to compare the properties of the alloys in the table.
4. Imagine you could have a bicycle made from any of the materials described in these pages. Explain which material you would choose, and why.

Key points

- Scientific findings are applied in society and industry.

Review 3.7

1. Use words from the list to complete the sentences.

physical	insulator
malleable	conductor
control	ductile
independent	alloy
brittle	

   a. Properties that you can observe or measure without changing a material are _____ properties.

   b. If electricity travels through a substance easily, the substance is a good _____ of electricity.

   c. If a substance can be hammered into different shapes, it is _____.

   d. If a substance can be pulled into wires, it is _____.

   e. The variable you change is the _____ variable.

   f. The variables you keep the same in a fair test are _____ variables.

   g. A mixture of a metal with one or more other elements is an _____.

   h. A substance that breaks easily when you hit it with a hammer is _____.

   i. If electricity cannot travel through a substance, the substance is an _____.

2. Complete the sentences using words from the list. You may use them once, more than once, or not at all.

   a good conductor of heat
   strong
   a good conductor of electricity
   sonorous

   a. Aluminium is used to make cooking pans because it is _____. [1]

   b. Copper is used in the cable of a lamp because it is _____. [1]

   c. Copper can be used to make bells because it is _____. [1]

   d. An iron alloy is used to make cars because it is _____. [1]

3. The picture shows some gold coins. They were made about 800 years ago in India.

   The list gives some properties of gold.

   It is a good conductor of electricity.
   It melts at 1063 °C.
   It is always shiny.
   It is a good conductor of heat.
   It is yellow.

   a. Give the one property in the list that shows that gold is in the solid state at 20 °C. [1]

   b. Give two properties in the list that are typical of all metals. [1]

   c. Give the one property in the list that best explains why gold was used to make coins. [1]

   d. Give the two properties in the list that best explain why gold is used to make connectors in some electrical devices. [1]

4. The table shows the properties of four elements. Each element is represented by a letter.

Element	Appearance at 20 °C	Does it conduct electricity?	Melting point (°C)
A	green	no	−101
B	shiny silver-coloured	yes	961
C	shiny grey	yes	1535
D	dull yellow	no	113

74

a. Give the letter of the element in the table that has the highest melting point. [1]

b. Give the letters of the elements in the table that are non-metals. Explain your choices. [2]

c. Give the letters of the two elements in the table that are likely to be good conductors of electricity. [1]

d. Give the letter of the one substance in the list that is likely to be brittle at 20 °C. [1]

5. A student wants to compare the strength of four wires, each made of a different metal or alloy. She sets up the apparatus below. She adds weights until the wires break.

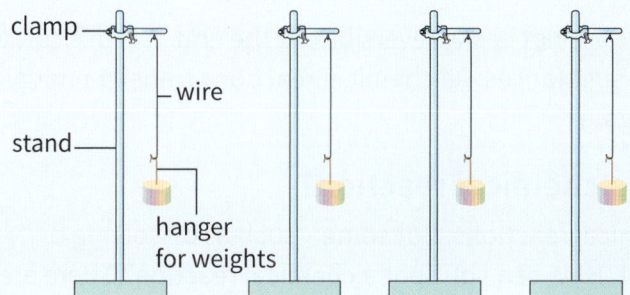

a. Identify the independent variable. [1]

b. Suggest two variables the student should control to make the test fair. [2]

c. The student's results are in the table.

Metal or alloy	Mass at which wire broke (kg)
copper	11
tungsten	8
stainless steel	5
aluminium alloy	3

  i. Name the two pure metals that the student used in the investigation. [1]

  ii. Name the metal or alloy that makes the strongest wires. [1]

  iii. Display the results on a bar chart. Use a copy of the axes at the top of the next column.

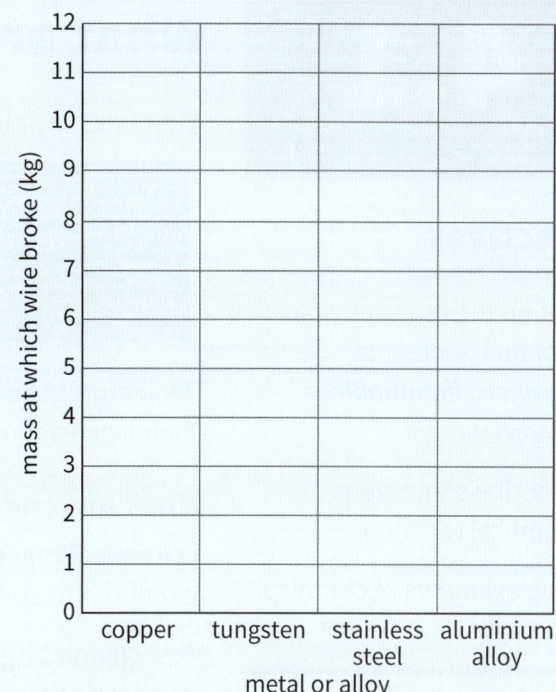

[3]

6. The bar chart below compares how well different metals conduct electricity.

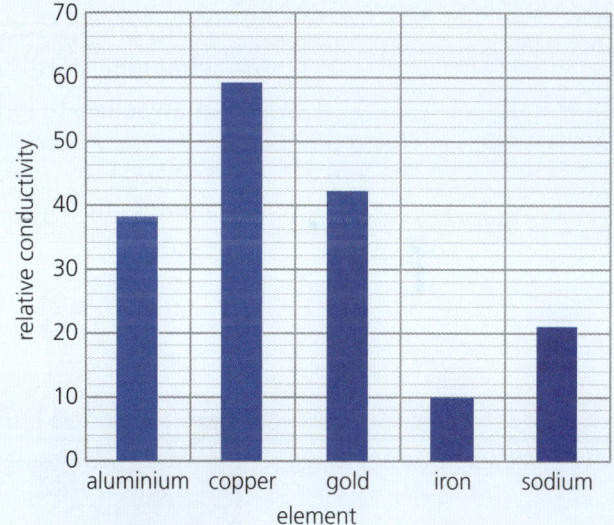

a. Explain why the data is presented in a bar chart, not in a line graph. [1]

b. Give the relative conductivity of gold. [1]

c. Give the relative conductivity of copper. [1]

d. List the metals in the bar chart in order of increasing relative conductivity (lowest first). [2]

e. Gold is a better conductor of electricity than aluminium, but long-distance electricity cables are made from aluminium, not gold. Suggest why. [1]

4.1 What are chemical reactions?

Objectives

- Define the terms *chemical reaction, reactants, products, flammable, hazard symbol*
- Describe some signs of chemical reactions
- Give examples of chemical reactions

What links the pictures below?

The pictures show chemical reactions, or places where they happen. Chemical reactions happen everywhere – even inside you.

What are chemical reactions?

A **chemical reaction** is a change that makes new substances. In a chemical reaction, the atoms are rearranged and join together differently.

Most chemical reactions are not easily reversible. At the end, it is difficult to get back to the starting substances. All chemical reactions transfer energy to or from the surroundings.

How can you spot a chemical reaction?

Many changes are chemical reactions. But some – such as dissolving or changing state – are not. How can you spot a chemical reaction? There are many clues to look out for. In a chemical reaction, you might:

▼ see huge flames...or tiny sparks

▼ notice a sweet smell...or a foul stink

▼ feel the substances get hotter...or colder

▼ hear a loud bang...or a gentle fizzing.

By the end of the reaction, what you see probably looks different to what you started with.

How does magnesium react?

Burning magnesium

Raj has a piece of shiny magnesium metal. Using tongs, he holds it in a flame. The magnesium burns brightly. Raj ends up with white ash.

Burning magnesium is a chemical reaction. The magnesium joins with oxygen from the air. The chemical reaction makes one new substance, magnesium oxide (the white ash).

Chemical reactions 1

▲ *Magnesium ribbon, before burning*

▲ *Burning magnesium*

▲ *Magnesium oxide*

In a chemical reaction:

- The starting substances are **reactants**.
- The substances that are made are **products**.

In the burning magnesium reaction, the reactants are magnesium and oxygen. There is one product – magnesium oxide. The reactants and product have different properties. For example, they look different.

🧪 Thinking and working scientifically

Working safely

Magnesium burns quickly. It is **flammable**. It has the **hazard symbol** shown on the right. A hazard symbol shows how a substance can cause harm. Hazard symbols are the same everywhere.

Raj's teacher reduces the risk from this hazard by looking after the roll of magnesium carefully. She keeps it away from flames.

Magnesium burns with a hot flame, so Raj holds it with tongs. The flame is bright white, and looking at it directly can damage eyesight. Raj reduces the risk from this hazard by looking at the flame through a narrow gap between his fingers.

▲ *'Flammable' hazard symbol*

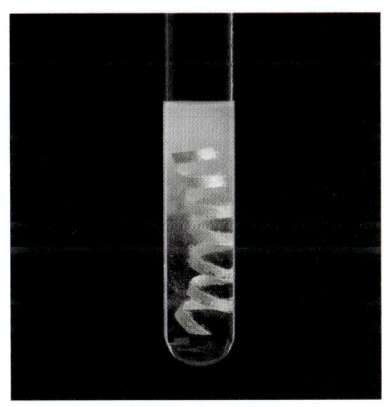
▲ *The chemical reaction of magnesium and hydrochloric acid*

Magnesium and acid

Raj adds another piece of magnesium to some hydrochloric acid. It bubbles. The bubbles are one of the products of the reaction – hydrogen gas. The properties of the reactants and products are different.

❓ Questions

1. Write the definition of *chemical reaction*.
2. Describe three observations you might make in a chemical reaction.
3. Name the reactants and product of the reaction of magnesium with oxygen.
4. **TWS** Describe two safety precautions when burning magnesium.
5. Magnesium reacts with iron oxide to make magnesium oxide and iron. Name the reactants and products of the reaction.

📖 Key points

- A chemical reaction is a change in which the atoms rearrange and join together differently to make new substances.
- The reactants and products of a chemical reaction have different properties.

4.2 Atoms in chemical reactions

A rocket blasts off. The flame is burning hydrogen.

Objectives

- Describe how atoms are rearranged in chemical reactions
- Explain why total mass does not change in chemical reactions
- Describe one way that science can have a global environmental impact

Atoms in chemical reactions

In the rocket flame, hydrogen and oxygen react together. The chemical reaction has one product – water. The diagram shows how the atoms rearrange and join together differently.

Two hydrogen molecules…   …react with…   …one oxygen molecule…   …to make…   …two water molecules.

There are the same number of atoms before and after the reaction:

- four hydrogen atoms
- two oxygen atoms.

In the rocket flame, millions and millions and millions of atoms rearrange and join together differently.

Science in context

Hydrogen cars and buses

Hydrogen does not only fuel rockets. It also fuels some buses and cars. In the engine or fuel cell, hydrogen reacts with oxygen. There is one harmless waste product, water.

Normal diesel buses make waste substances that harm health. They also make carbon dioxide. As you know, carbon dioxide is a greenhouse gas. It makes the Earth's surface hotter, which causes climate change.

Using hydrogen instead of diesel to fuel buses could reduce the amounts of harmful substances and greenhouse gases. This might have a positive impact on the environment, all over the world.

But hydrogen fuel is not perfect:

- Hydrogen is difficult to store.
- There are few hydrogen filling stations.

▲ *This bus uses hydrogen fuel.*

Chemical reactions 1

Mass in chemical reactions

The numbers of atoms of each element do not change in a chemical reaction. This means that the mass does not change. In any chemical reaction, the total mass of products is equal to the total mass of reactants.

Halim weighs a piece of magnesium, and heats it in a crucible. While heating, he lifts the lid so that air can get in. The magnesium reacts with oxygen from the air to make magnesium oxide. After heating, Halim weighs the product.

Mass of magnesium at start = 0.24 g
Mass of magnesium oxide at end = 0.40 g

▲ Heating magnesium in a crucible

The product mass is greater than the magnesium mass. This is because magnesium reacts with oxygen from the air. Oxygen has its own mass.

The mass of magnesium and oxygen at the start is equal to the mass of magnesium oxide made. Halim calculates that 0.16 g of oxygen joins to the magnesium.

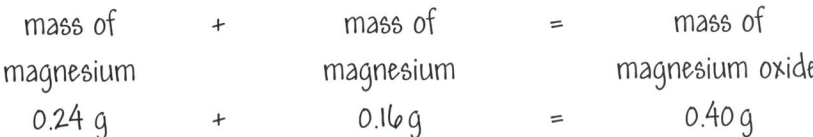

The diagram shows some atoms before and after the reaction of magnesium and oxygen. There are the same number of magnesium atoms before and after. There are the same number of oxygen atoms before and after. This explains why the total mass is the same before and after.

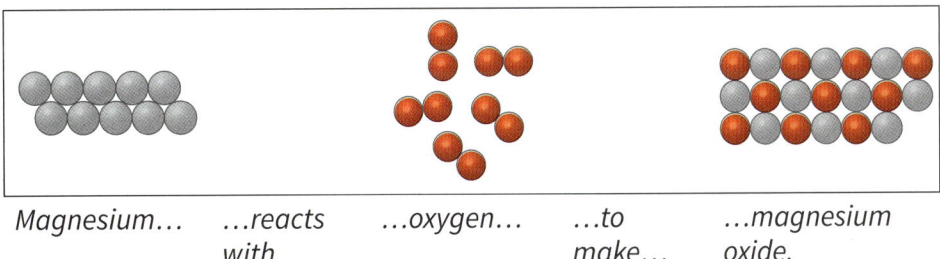

Magnesium…   …reacts with…   …oxygen…   …to make…   …magnesium oxide.

❓ Questions

1. Describe what happens to the atoms in a chemical reaction.
2. Explain why the total mass of reactants is equal to the total mass of products in a chemical reaction.
3. In a chemical reaction, 32 g of sulfur reacts with 32 g of oxygen to make sulfur dioxide. Calculate the mass of sulfur dioxide made.
4. Suggest a positive impact on the environment if all buses used hydrogen fuel instead of diesel.

📖 Key points

In a chemical reaction:
- atoms are rearranged and join together differently
- the total mass of products is equal to the total mass of reactants.

Thinking and working scientifically 4.3

Objective

- Describe some stages in a scientific enquiry

Investigating a chemical reaction

Big pieces of iron do not burn. But thin strands of iron, called iron wool, burn well. Zac wants to investigate this chemical reaction.

▲ Iron wool burns well.

Planning the investigation

Making a hypothesis

Zac knows that when magnesium burns, it reacts with oxygen to make magnesium oxide. He thinks that iron might react in a similar way. He uses this scientific knowledge to make a hypothesis. As you know, a hypothesis is a possible explanation that is based on evidence, and can be tested further.

My hypothesis – If I heat iron wool in air, the iron will react with oxygen. This will make a compound, iron oxide.

Zac plans to collect evidence to test his hypothesis. He will burn some iron in air. If the mass of product is greater than the mass of iron, this will be evidence that iron reacts with oxygen from the air.

Plan and prediction

Zac asks his teacher for help. The teacher draws some apparatus.

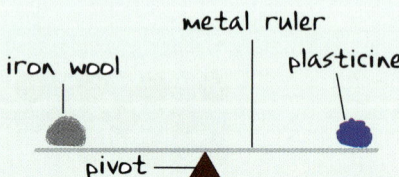

Zac plans to heat the iron wool with a hot flame. The iron will burn as it reacts with oxygen from the air.

Zac knows that iron atoms have mass. Oxygen atoms have mass too. He makes a prediction.

If I heat iron in air, iron atoms and oxygen atoms will join together. The mass of iron oxide I make will be greater than the mass of iron I start with. So the ruler will tilt like this:

Chemical reactions 1

Carrying out the investigation

Equipment

Zac thinks about the equipment to use. He knows that a wooden ruler might catch fire, so he uses a metal ruler.

Safety

Zac thinks about the hazards. He decides how to work safely:

- Iron wool can cut skin, so wear gloves.
- Iron wool catches fire easily, so keep spare iron wool in a jar with a lid.

He also wears eye protection.

Analysing the evidence

Making a conclusion

Zac heats the iron wool for one minute. The ruler tilts over like this. His prediction was correct.

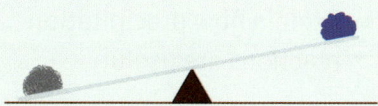

Zac writes a conclusion:

Conclusion – My prediction was correct. The iron wool end of the ruler tilted down.

Zac's teacher asks him to improve the conclusion. She tells him to use scientific knowledge to explain his results. Zac adds the sentences below:

When I heated the iron, iron atoms joined with oxygen atoms from the air. This made a compound, iron oxide. The mass of iron oxide was greater than the mass of the iron at the start. So the iron oxide end tilted down.

Making improvements

Zac wants to improve his investigation. He does the experiment again. This time, he makes some measurements.

Mass of iron before heating = 4.1 g
Mass of product (iron oxide) after heating = 5.6 g

Zac knows that, in a chemical reaction, the total mass of reactants is equal to the total mass of products. He calculates the mass of oxygen that joins to the iron.

mass of iron + mass of oxygen = mass of iron oxide
  4.1 g   + mass of oxygen =   5.6 g

mass of oxygen = mass of iron oxide – mass of iron
         =   5.6 g   –   4.1 g
         =   1.5 g

Key points

When planning an investigation, use scientific knowledge to:

- make a hypothesis and prediction
- choose equipment
- work safely
- make a conclusion
- suggest improvements.

Questions

1. Write the definition for *hypothesis*.
2. a. List two hazards in Zac's investigation.
   b. Describe what Zac did to work safely.
3. Write a list of the steps in Zac's investigation.

81

4.4 Precipitation reactions

Zamira has some potassium iodide solution. It is colourless. She adds some lead nitrate solution, which is also colourless. She watches as tiny pieces of a yellow solid are made.

Objectives

- Define the terms *precipitation reaction* and *precipitate*
- Explain why precipitates form
- Explain how precipitation reactions are useful
- Give the meanings of some hazard symbols

Thinking and working scientifically

Working safely

Lead nitrate has these hazard symbols. Zamira works safely by wearing gloves and eye protection. She does not pour the solution down the sink.

Symbol	The substance…
(flame over circle)	…helps other substances to burn.
(health hazard)	…is a health hazard.
(corrosion)	…can burn skin and eyes.
(environment)	…damages things that live in water.

▲ Adding lead nitrate to potassium iodide solution makes a yellow precipitate.

Different types of chemical reaction

There are different types of chemical reaction. These include:

- Burning, or combustion reactions. In a **combustion reaction**, a substance reacts quickly with oxygen and gives out light and heat. The burning reaction of magnesium is a combustion reaction.
- Precipitation reactions, like Zamira's reaction. Read on to find out more.

In all types of chemical reaction atoms are rearranged and join together differently to make new substance.

What is a precipitation reaction?

In a **precipitation reaction**, two reactants in solution react to make a precipitate. A **precipitate** is a suspension of tiny solid particles in a liquid or solution. A precipitate forms in a chemical reaction when one product is insoluble. As you know, an insoluble substance is one that does not dissolve.

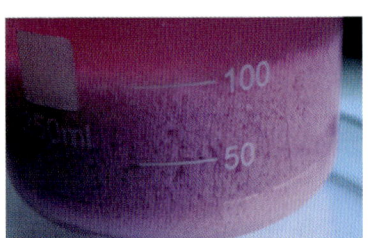

▲ This precipitate is cobalt carbonate. It formed in the chemical reaction of cobalt sulfate solution and sodium hydrogencarbonate solution.

Chemical reactions 1

The reactants and products in a chemical reaction have different chemical and physical properties. In precipitation reactions, colour and solubility are different for the reactants and products.

This is what happens in the reaction of potassium iodide solution with lead nitrate solution:

potassium iodide… (colourless solution)	…reacts with…	…lead nitrate… (colourless solution)	…to make…	lead iodide… (yellow precipitate)	…and…	…potassium nitrate (colourless solution)

This is not the only precipitation reaction. Other pairs of soluble substances react to make a precipitate if one product is insoluble. As you know, a soluble substance is one that dissolves.

Using precipitation reactions

You can use precipitation reactions to help identify substances. Advik has a solution. He does not know what it is. He adds a few drops of sodium hydroxide solution. A blue precipitate forms. This shows that the substance in the solution is a copper compound.

This is what happens in Advik's experiment.

copper chloride… (solution)	…reacts with…	…sodium hydroxide… (solution)	…to make…	…copper hydroxide… (blue precipitate)	…and…	…sodium chloride (colourless solution)

Different metal compounds make precipitates of different colours.

◂ Iron compounds make dark green or brown precipitates. Copper compounds make blue precipitates. Nickel compounds make light green precipitates.

Key points

- In a precipitation reaction, two solutions react to make a precipitate.
- A precipitation reaction occurs when two soluble reactants form an insoluble product.
- Precipitation reactions help to identify unknown substances.

Questions

1. Write the definitions for *precipitation reaction*, *soluble*, and *insoluble*.

2. A substance has this hazard symbol. Give the meaning of the symbol and suggest how to work safely with the substance.

3. Advik has a solution of an unknown compound. He adds a few drops of sodium hydroxide solution. A brown precipitate forms. Name the metal whose atoms are in the unknown compound.

4.5 Corrosion reactions

Objectives

- Define the terms *corrosion* and *chemical properties*
- Name the reactants and product in the corrosion of iron
- Name an alloy that has different chemical properties to the elements that are in it

Many chemical reactions are useful. Combustion reactions release energy to cook food. Precipitation reactions help to identify substances. Other chemical reactions make medicines and fertilisers. Chemical reactions even keep living things living.

But some chemical reactions are not useful. Read on to find out more.

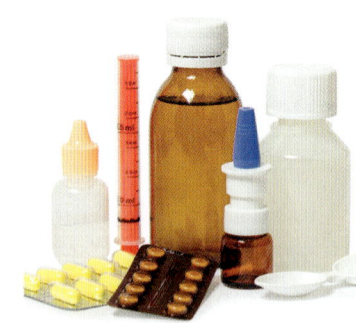

▲ *The car and boat are rusty.*

What are corrosion reactions?

The car and the boat are made from steel, an alloy of iron. They are covered in rust. Rust is formed in a **corrosion** reaction. Corrosion is a chemical reaction that happens on the surface of a metal. It may damage metal objects.

The corrosion of iron is called rusting. When iron rusts, it reacts with oxygen and water. The oxygen comes from the air. The water may be in the air as gas or in its liquid state. Rusting happens slowly, over weeks and months.

The product of the rusting reaction is hydrated iron oxide. This is iron oxide with water loosely joined to it.

iron…	…reacts with…	…oxygen…	…and…	…water…	…to make…	…hydrated iron oxide
(shiny solid)		(colourless gas)		(colourless gas or liquid)		(flaky brown solid)

The rust forms on the surface of the iron. It is soft and crumbly. It easily comes off the surface of the metal. This leaves more iron exposed and ready to rust.

What are chemical properties?

Magnesium corrodes quickly. Iron corrodes slowly. Some metals, such as gold, do not corrode at all. Whether or not a substance corrodes is one of its chemical properties. The **chemical properties** of a substance describe its chemical reactions, such as whether it burns, corrodes, or reacts to make a precipitate.

Chemical properties are different to physical properties. You can only find out about chemical properties in chemical reactions. But you can observe physical properties without changing the substance.

How can we prevent corrosion?

As you know, steel is mainly iron. Most of the metal things we use are made from steel. So iron corrosion – rusting – is an expensive problem.

Coating with paint, oil, or zinc

Oxygen and water make iron rust. You can prevent rusting by keeping oxygen and water away from iron. Scientists do this by coating iron and steel objects with:

- paint (on bikes, cars, and boats)
- oil or grease (in engines)
- a metal that does not easily corrode, such as zinc.

▲ *The steel bucket is coated with zinc.*

Alloys

As you know, an alloy is a mixture of a metal with another element. Every alloy has physical properties that are different to the elements that are mixed to make it.

Some alloys also have different chemical properties to their elements. Stainless steel is an alloy of iron, mixed with about 10% chromium. Stainless steel does not react with oxygen and water, so it does not rust.

Stainless steel is more expensive than other types of steel. It is worth the extra cost for things that must not rust, but cannot be covered in paint or oil.

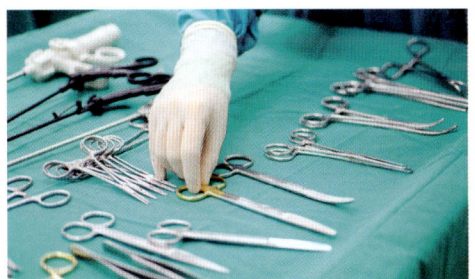

▲ *Surgeons use stainless steel tools in operations.*

▲ *This concert hall is made from stainless steel.*

📖 Key points

- Corrosion is a chemical reaction that happens on the surface of a metal.
- Chemical properties describe the chemical reactions of a substance.
- Rusting is the corrosion of iron.
- Stainless steel is an alloy of iron. It has different chemical properties to the elements that are mixed to make it. It does not rust.

❓ Questions

1. Write the definitions for *corrosion* and *chemical properties*.
2. Name the reactants and product when iron rusts.
3. Explain why rusting is a problem.
4. Describe three ways of preventing rusting, and explain how they work.
5. Give one chemical property that is different for iron and one of its alloys, stainless steel.

Review 4.6

1. The pictures show some hazard symbols. Write the meaning of each hazard symbol. Choose from the list below. [5]

   burns easily
   can burn skin and eyes
   damages things that live in water
   helps other things to burn
   is a health hazard

   a.
   b.
   c.
   d.
   e.

2. Copy and complete the sentences. Choose words and phrases from the list below. [6]

   chemical reaction
   combustion
   corrosion
   products
   reactants

   a. A change that makes new substances is a _____ _____.
   b. The starting substances in a chemical reaction are _____.
   c. The substances that are made in a chemical reaction are _____.
   d. Burning is also called _____.
   e. Rusting is also called _____.

3. Hydrogen and chlorine react to make hydrogen chloride. The diagram shows the atoms in the reaction.

     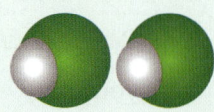

   Hydrogen… …reacts with… …chlorine …to make… …hydrogen chloride

   a. Name the reactants. [1]
   b. Name the product. [1]
   c. Give the colour of the chlorine atoms in the diagram above.
   d. Give the number of chlorine atoms shown in the reactants. [1]
   e. Give the number of chlorine atoms shown in the product. [1]
   f. Explain why the total mass of reactants is equal to the total mass of product in the reaction shown above. Use the diagram to help you explain. [2]

4. A student carries out some chemical reactions. Write down the type of each chemical reaction. Choose from the list below. Use each word once, more than once, or not at all. [4]

   combustion
   corrosion
   precipitation

	What the student does	Observations
a.	Adds sodium chloride solution to silver nitrate solution.	Makes a white mixture that you cannot see through.
b.	Leaves an iron nail in a mixture of water and oxygen for a week.	Brown flakes form on the surface of the nail.
c.	Sets fire to a piece of wood.	Produces orange flames.
d.	Adds sodium hydroxide solution to copper nitrate solution.	Makes a blue mixture that you cannot see through.

5. The sign in the picture is made from iron. It is outside a shop. It was painted a long time ago.

   a. Name the two substances that reacted with the sign to make it go brown. [2]

   b. Name the type of chemical reaction that made the sign go brown. [1]

   c. The shop owners get a new iron sign. Suggest what they could do to stop the sign going brown. [1]

6. In a chemical reaction, the total mass of reactants is equal to the total mass of products.

   a. 4 g of hydrogen reacts with 32 g of oxygen to make water only. Calculate the mass of water that is made. [2]

   b. 1.2 g of carbon reacts with 3.2 g of oxygen to make carbon dioxide only. Calculate the mass of carbon dioxide that is made. [2]

   c. In a chemical reaction, 56 g of iron reacts with sulfur to make 88 g of iron sulfide. Calculate the mass of sulfur that reacts. [2]

   d. In a chemical reaction, 32 g of sulfur reacts with oxygen to make 64 g of sulfur dioxide. Calculate the mass of oxygen that reacts. [2]

7. Sahira wants to investigate mass changes when she burns metals in air.

   a. Sahira writes down two possible questions to investigate.
   ~ how do the masses of all metals change when they burn?
   ~ what is the mass change of magnesium when it burns?
   Sahira's teacher tells her to investigate the second question above. Suggest why. [1]

   b. Sahira draws the apparatus for investigating the mass change of magnesium when it burns to make magnesium oxide.

   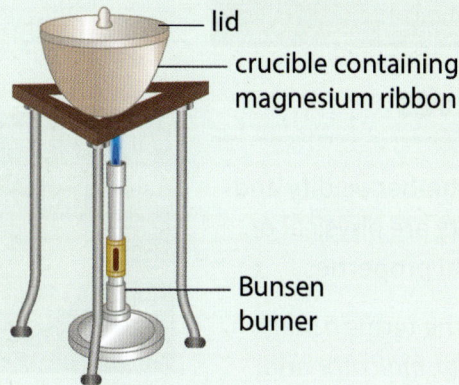

   Sahira must open the lid a few times as she heats. Suggest why. [1]

   c. Sahira writes down some hazards of the investigation. Copy and complete the table to show how she can work safely.

Hazard	How I can work safely
magnesium is highly flammable	
burning magnesium has a very bright flame	
the apparatus gets very hot	

   d. Sahira writes down her results.
   Mass of crucible + lid = 32.00 g
   Mass of magnesium = 0.24 g
   Mass of crucible + lid + magnesium oxide = 32.40 g

   i. Calculate the mass of magnesium oxide made. [2]

   ii. Calculate the mass of oxygen that joined with the magnesium in Sahira's experiment. [2]

   e. Sahira writes the conclusion below.
   In my investigation, the mass increased.
   Add a scientific explanation to improve Sahira's conclusion. [2]

5.1 Acids and alkalis

Why do lemons, limes, and vomit taste sour? They all contain acids. They are acidic. Toothpaste, soap, and baking soda are not acidic. They are alkaline.

Objectives

- State whether acidity and alkalinity are physical or chemical properties
- Define the terms *acidity, alkalinity, indicator* and *neutral*
- Describe how to use indicators to detect acidic and alkaline solutions

▲ Lemon juice is acidic.

▲ Toothpaste and soap are alkaline.

Some substances are **neutral**. They are not acidic or alkaline. Pure water is neutral.

The **acidity** or **alkalinity** of a substance is a chemical property. As you know, chemical properties of a substance describe its chemical reactions.

How can you detect acids and alkalis?

Latif has two beakers. One contains an acidic solution. The other contains an alkaline solution. How can he find out which is which?

Latif uses an indicator. An **indicator** is a solution of a dye that changes colour in acidic and alkaline solutions.

You can make indicators from plants. The table shows the colours of some plant indicators in acidic and alkaline solutions.

▲ Which solution is acidic, and which is alkaline?

▲ One indicator is litmus. Litmus solution is red in acid. It is blue in an alkaline solution.

▲ Paper is soaked in litmus to make litmus paper. Blue litmus paper turns red in acid. Red litmus paper turns blue in an alkaline solution.

▲ Red cabbage indicator is red in acidic solutions. It is dark blue, yellow, or green in alkaline solutions.

Juice from...	Colour in acidic solution	Colour in alkaline solution
red cabbage	red	yellow / green / blue
hibiscus flower	dark pink / red	dark green
beetroot	red / purple	yellow

88

Thinking and working scientifically

Choosing and using apparatus

Making an indicator

Femi makes hibiscus flower indicator at home. She adds hibiscus flowers to warm water in a cooking pot. She pours the mixture through a colander and collects the solution in a cup.

Adamma makes hibiscus flower indicator in a science lab. She adds hibiscus flowers to water in a beaker. She heats the mixture with a Bunsen burner. She pours the mixture through filter paper and collects the solution in a conical flask.

▲ *Hibiscus flowers make a good acid–alkali indicator.*

Both Femi and Adamma have chosen suitable apparatus, since both made an indicator that works.

▲ *Femi's apparatus*

▲ *Adamma's apparatus*

Testing the indicator

Femi places some acid on a green plate. She adds her indicator. She does the same test with an alkali. She writes down the colours she sees.

Adamma pours some acid into a test tube. She adds some indicator. She does the same test with an alkali. She writes down the colours.

In this part of the experiment, Femi's apparatus is not suitable. She cannot see the colours on the green plate. Adamma made a better choice of apparatus.

Questions

1. State whether acidity and alkalinity are physical or chemical properties.
2. Write the definitions for *indicator* and *neutral*.
3. A student adds red cabbage indicator to an acidic solution. Predict the colour of the mixture.
4. A student has a solution, X. He adds hibiscus indicator. The solution is now dark green. The student adds beetroot indicator to a separate sample of solution X. Predict its colour.

Key points

- The acidity or alkalinity of a substance is a chemical property.
- Some substances are not acidic or alkaline – they are neutral.
- Indicators show whether a substance is acidic or alkaline.

5.2 The pH scale

Some acidic substances are good to eat or drink. But others are corrosive – they would burn your mouth. How can we measure acidity safely?

Objectives

- Define the terms *acid* and *alkali* in terms of pH
- Classify solutions as acidic, alkaline, or neutral
- Describe how to measure pH
- Describe some applications of science

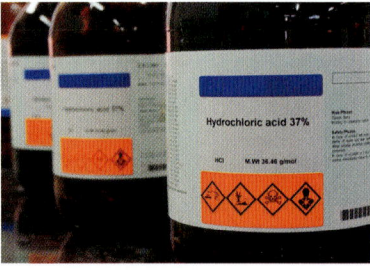 

◄ Hydrochloric acid is corrosive. Another acidic solution, tea, is safe to drink.

The pH scale

The pH scale shows how acidic or alkaline a solution is, or whether it is neutral:

- An acid has a pH less than 7.0. The lower the pH, the more acidic the solution.
- An alkali has a pH more than 7.0. The higher the pH, the more alkaline the solution.
- A neutral solution has a pH of 7.0.

Thinking and working scientifically

Classifying substances

Salma uses the Internet to find the pH of some solutions. Her data are in the table. She uses the pH values to classify some of the solutions as acidic, alkaline, or neutral.

Solution	pH	Is the solution acidic, alkaline, or neutral?
black coffee	5.5	acidic
blood	7.4	alkaline
drain cleaner	13.0	
lemon juice	2.3	
milk	6.6	
orange juice	3.2	
sodium hydroxide	14.0	
sulfuric acid	1.0	
tea	6.0	
vinegar	2.8	
pure water	7.0	neutral

The data in the table are secondary data. **Secondary data** are data that are collected by someone else – in this case, not by Salma.

Acids and alkalis

How do you measure pH?

Litmus is red in all acidic solutions. It is blue in all alkaline solutions. You cannot use litmus to measure pH. Instead, you need universal indicator. **Universal indicator** is a mixture of dyes that changes colour to show how acidic or alkaline a substance is.

To find the pH of a solution, add a few drops of universal indicator. Match its colour to the colour chart. You can also use universal indicator paper.

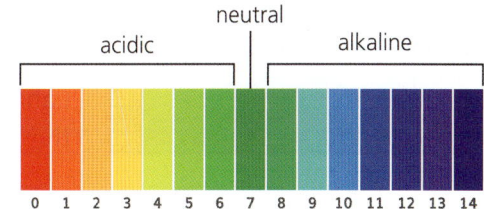
▲ *Universal indicator colours*

You can also use an electronic device, called a pH probe, to measure pH.

▲ *Universal indicator paper and colour chart.*

▲ *The probe shows that the soil pH is 6.5.*

Science in context

Using and discovering sulfuric acid

Sulfuric acid is very important to humans. It is used to make fertilisers and insecticides to help grow food crops. It is also used to make detergents, dyes, medicines, paints, and batteries. Worldwide, factories make over 240 million tonnes of sulfuric acid each year.

▲ *A sulfuric acid factory*

▲ *Jabir Ibn Hayyan*

Discovering sulfuric acid

Around 1200 years ago, Jabir Ibn Hayyan worked in what is now Iraq. He discovered how to make sulfuric acid (as well as nitric acid and hydrochloric acid). He was probably the first person to use the word 'alkali'. Jabir wrote more than 100 books about his findings. Scientists used these for many years.

Questions

1. Give the pH of a neutral solution.
2. Look at the table on the left.
   a. Write two lists – one list of acidic solutions and one list of alkaline solutions. Then name the one neutral solution
   b. Give the name of the most acidic solution in the table.
3. Sadiq has three acidic solutions. Describe how he can use universal indicator to find out which solution is most acidic, and which is least acidic.

Key points

- An acidic solution has a pH less than 7 and an alkaline solution has a pH greater than 7.
- pH is measured with universal indicator or a pH probe.
- Sulfuric acid has many uses.

5.3 Neutralisation reactions

Lahan has half a lemon. He puts baking soda on the lemon. It fizzes. Why?

Objectives

- Define the term *neutralisation*
- Describe how to neutralise an acid
- Describe some applications of science

What is neutralisation?

The fizzing shows that there is a chemical reaction. There are two reactants:

- citric acid (in the lemon juice)
- sodium hydrogencarbonate (baking soda).

▲ *Baking soda and lemon*

One of the products, carbon dioxide, forms as a gas. This makes bubbles.

The chemical reaction is an example of a neutralisation reaction. **Neutralisation** is a type of chemical reaction in which an alkali reacts with an acid and the pH gets closer to 7.

Neutralisation in the lab

You can do neutralisation reactions in the lab. Hasina has 100 cm^3 of hydrochloric acid. She also has sodium hydroxide solution, which is alkaline. She wants to find the volume of alkali that neutralises the acid. She follows the steps below.

Step 1

Add universal indicator to the acid.

Step 2

Add sodium hydroxide (an alkali) to the acid-indicator mixture.

Step 3

Continue to add the alkali to the mixture, until its colour is green.

▲ *The mixture is red, showing that its pH is 1. The hydrochloric acid is acidic.*

▲ *The mixture is orange. Its pH is 3. The alkali has neutralised some of the acid.*

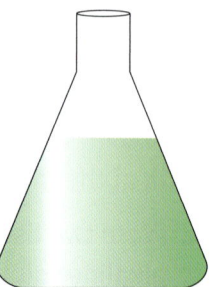

▲ *The mixture is green. Its pH is 7. The alkali has neutralised all the acid.*

Science in context

Using neutralisation reactions

Soil pH

Some soils are more acidic or alkaline than others. Every plant has its preferred soil pH. Carrots grow well in soil of pH 6.0. Tea grows best in soil of pH 4.0 to 5.5. Date palms prefer pH 6.5 to 8.0.

The table shows the preferred soil pH of some food plants.

Vegetable	Preferred soil pH
cabbage	6.0–7.0
onion	6.0–6.5
maize	5.5–7.0
sweet potato	5.0–6.0
tomato	5.5–6.8

Zainab takes some soil from her garden. She mixes it with pure water. She adds universal indicator. The indicator shows that the soil is acidic – its pH is 5.0. The soil is suitable for sweet potatoes.

Zainab also wants to grow cabbage and onions. But they grow best on soils of higher pH. Zainab adds an alkali to the soil in one part of the garden. This neutralises some of the acid. The soil pH increases. Zainab can now grow cabbage and onions.

Treating stomach ache

If your stomach produces excess acid, you may get stomach ache. Indigestion tablets contain an alkali or a carbonate. This neutralises the extra acid. See Unit 5.4 to find out more.

▲ Sweet potatoes grow best in slightly acidic soil of pH 5.0 to 6.0.

▲ Date palms grow best in soil of pH 6.5 to 8.0.

Questions

1. Write the definition for *neutralisation*.
2. Ola has a solution of pH 9. Explain whether he should add acid or alkali to neutralise the solution.
3. Use the table above to help you answer this question. Ejaz measures the soil pH on his farm. The pH is 8.0. Ejaz wants to grow onions. Explain whether he should add an acid or an alkali to the soil.

Key points

- Neutralisation is a type of chemical reaction in which an alkali reacts with an acid and the pH gets closer to 7.
- Neutralisation changes soil pH and treats stomach ache.

Thinking and working scientifically

5.4

Objective

- Describe how to plan, carry out, and analyse a scientific enquiry

Investigating neutralisation

Kali has stomach ache. He takes an indigestion tablet. The tablet contains a substance that neutralises extra acid in his stomach. In a few minutes, Kali feels better.

Kali saw three different types of indigestion tablet in the pharmacy. He wants to find out which type of tablet is best. He decides to investigate.

Planning the investigation

Suggesting ideas to test

Kali thinks of questions to investigate:

- Which type of tablet is best?
- Which type of tablet makes the pH increase most when added to acid?

The first question cannot be investigated using science alone. Kali decides to investigate the second question. He does not make a hypothesis or a prediction, since he has no scientific evidence to base it on.

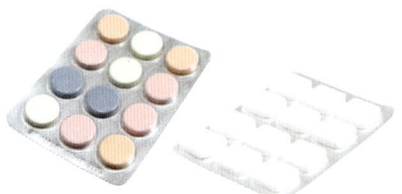

▲ Different indigestion tablets

Considering variables

Kali thinks about the variables. He knows that changing one variable will affect another variable. This means he can answer the question by doing a fair test. He lists the variables:

- Independent variable (what I will change) – type of tablet
- Dependent variable (what I will measure) – pH change when I add a tablet to the stomach acid
- Control variables (what I will keep the same) – type of acid, amount of acid.

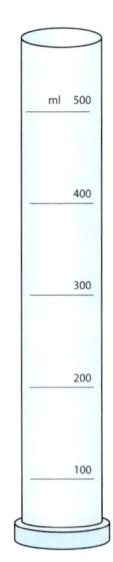

▲ A measuring cylinder

Carrying out the investigation

Choosing equipment

Kali chooses hydrochloric acid to represent stomach acid. He uses a measuring cylinder to measure equal volumes of acid for the three tests. A measuring cylinder measures smaller differences in volume than a beaker.

Working safely

▲ Acid may be corrosive.

The acid in the experiment is corrosive. This means that the acid can burn skin or eyes. Kali wears eye protection. He is careful not to get acid on his skin.

94

Collecting evidence

Kali pours the acid into a conical flask. He adds universal indicator. He uses a colour chart to find the acid pH.

Next, Kali adds a tablet to the acid. He waits for the bubbling to stop, which shows that the reaction has finished. He measures the pH of the mixture of products. He records the new pH in a table.

Finally, Kali repeats the experiment for the other two types of tablet. His results are in the table.

▲ *A conical flask and universal indicator*

Type of tablet	pH before adding tablet	pH after adding tablet	Change in pH
A	1	4	3
B	1	1	0
C	1	5	4

Analysing the evidence

Making a conclusion

Kali looks at the results in the table. He thinks about his question. He writes a conclusion that includes a scientific explanation:

Tablet C caused the greatest increase in pH. This means that tablet C neutralised the most acid.

Making improvements

Kali was surprised that the pH did not change with tablet B. Had he made a mistake? He repeated the test for tablet B three more times. His new results are in the table.

Tablet B test number	pH before adding tablet	pH after adding tablet	Change in pH
1	1	1	0
2	1	4	3
3	1	4	3
4	1	4	3

Kali noticed that the result for test 1 is anomalous – it is different to the others. He decided to ignore this result. Tablet B changes the pH by 3. This does not affect Kali's conclusion, since the pH change is still greatest with tablet C.

Kali's teacher asks him to repeat the tests for tablets A and C. This will make all the results more reliable.

> **Key points**
> - An investigation can involve considering variables, choosing equipment, collecting evidence, making a conclusion, and making improvements.

> **Questions**
> 1. Write the definitions for *independent variable* and *dependent variable*.
> 2. Explain why Kali kept the control variables constant in his investigation.
> 3. Explain why Kali uses a measuring cylinder to measure equal volumes of acid, not a beaker.

Science in context 5.5

Acid rain

The Taj Mahal in India is turning yellow. Its white marble is being damaged. Why? Can the damage be stopped?

Objective

- Evaluate issues using science

What is acid rain?

Near the Taj Mahal is a big city. More and more people in the city drive cars. Lorries bring things that people need. And next to the Taj Mahal, piles of rubbish burn.

There are factories near the Taj Mahal, too. Some make glass. Others process leather. One makes products from oil.

▲ Some factories make acidic waste gases.

▲ Cars and lorries make acidic gases.

The cars, lorries, factories, and burning rubbish make acidic gases. The gases mix with the air. They dissolve in rain, making it acidic.

When acid rain falls on the Taj Mahal, there are chemical reactions. The acid reacts with the marble to make products such as carbon dioxide, water, and substances in solution. The marble is damaged for ever.

One of the chemical reactions has two reactants. They react like this:

▲ Marble reacts with acid.

Nitric acid… (colourless solution in acid rain)	…reacts with…	…calcium carbonate (solid white marble)	…to make…	…calcium nitrate… (colourless solution)	…and…	…water… (liquid)	…and…	…carbon dioxide (colourless gas)

How can the damage be prevented?

Possible action

Science knowledge suggests how to prevent damage to the Taj Mahal.

- Factories can neutralise their acidic waste.
- People can drive electric or hydrogen cars (or walk or cycle).

Some prevention methods are easier than others. Some methods are more effective. Some methods are cheaper.

Deciding what to do

The government used science to evaluate the possible methods. They also considered other issues, such as the jobs that factories provide.

The government made a decision. They told factories to neutralise their waste, or move away. They banned petrol and diesel vehicles in the area.

Some people obeyed the rules, but others did not. In 2021, acid rain continued to damage the Taj Mahal.

Other acid rain problems

Acid rain does not only damage buildings. In some countries, power stations burn coal to make electricity. Burning coal makes waste products, including acidic gases.

Some power stations neutralise the acidic gases. Some power stations send acidic gases into the air. The gases dissolve in rainwater, making it acidic.

▲ *Acid rain damages trees.*

Acid rain damages trees. It makes lakes acidic, so some plants and animals cannot live in them. Acid rain also damages some types of rock. Scientists think that this may have caused a landslide in China.

Taking action

Governments use science to help them decide what to do. They may prevent acid rain by:

- telling power stations to remove acidic waste gases
- using cleaner methods to make electricity, such as wind turbines or solar cells.

▲ *Adding an alkali to an acidic lake*

If a lake is acidic, an alkaline substance can be added. This neutralises the extra acid.

Every possible action has advantages and disadvantages. Governments need to evaluate these before deciding what to do.

🔑 Key points

- Acid rain damages buildings and trees and makes lakes acidic.
- Science knowledge helps to evaluate some problems.

❓ Questions

1. Describe three problems caused by acid rain.
2. Do you think that it is better to prevent acid rain, or to solve the problems it causes? Give reasons for your answer.

5.6 Gas products of acid reactions

If you add an indigestion tablet to water, it may bubble. Why? What is the gas in the bubbles?

Objectives

- Name the gas products of the reactions of acids with metals and with carbonates
- Describe how to test for hydrogen, carbon dioxide, and oxygen gases

Acids and carbonates

Some indigestion tablets contain calcium carbonate. When you add calcium carbonate to an acid, there is a chemical reaction. The calcium carbonate neutralises the acid. One of the products is carbon dioxide gas. As you know, the formula of carbon dioxide is CO_2. There are two other products.

For example, hydrochloric acid reacts with calcium carbonate. The reaction is like the reaction of acid rain with marble, in Unit 5.4.

▲ An indigestion tablet bubbles in water.

Hydrochloric acid…	…reacts with…	…calcium carbonate…	…to make…	…calcium chloride…	…water…	…and…	…carbon dioxide
(colourless solution)		(colourless solution)		(colourless solution)	(liquid)		(colourless gas)

Testing for carbon dioxide gas

You can test for carbon dioxide gas like this:

- Bubble the gas through limewater solution.
- If the limewater solution turns milky or cloudy, carbon dioxide is present.

▲ Testing for carbon dioxide.

Acids and metals

Some metal elements react with acids. For example, hydrochloric acid reacts with magnesium. One of the products – hydrogen – is formed as a gas, H_2:

magnesium…	…reacts with…	…hydrochloric acid…	…to make…	…magnesium chloride…	…and… …hydrogen
(colourless solution)		(colourless solution)		(colourless solution)	(colourless gas)

You can read more about the reactions of acids with metals in Chapter 9.

Acids and alkalis

Testing for hydrogen gas

You can test for hydrogen gas like this:

- Collect the gas by holding an empty test tube above the reaction test tube.
- Light a splint.
- Place the lighted splint in the gas test tube.
- Listen. If there is a squeaky pop, the gas is hydrogen.

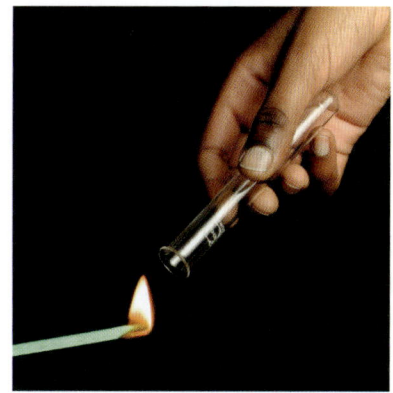

▲ *If you hear a squeaky pop, hydrogen is present.*

Testing for oxygen gas

You can also test for oxygen gas, O_2. This is what to do:

- Collect the gas by holding an empty test tube above the reaction test tube.
- Light a splint and blow it out.
- Place the glowing splint in the gas.
- If the splint relights, the gas is oxygen.

▲ *If a glowing splint relights, the gas is oxygen.*

Questions

1. Name the gas product in these reactions:
   a. The reaction of magnesium with hydrochloric acid.
   b. The reaction of calcium carbonate with hydrochloric acid.
2. Describe how to test for carbon dioxide gas. Include what to do and what you would see.
3. Describe how to test for oxygen gas. Include what to do and what you would see.
4. Meera puts a lighted splint in a test tube of gas. It goes out with a squeaky pop. Name the gas in the test tube.

Key points

- Acids react with carbonates to make carbon dioxide gas.
- Acids react with some metals to make hydrogen gas.
- Carbon dioxide makes limewater milky.
- Hydrogen puts out a lighted splint with a squeaky pop.
- Oxygen relights a glowing splint.

99

Review 5.7

Questions

1. Complete the sentences using phrases from the list. You may use them once, more than once, or not at all.

   | more than | equal to | less than |

   The pH of an acid is _____ 7. The pH of an alkali is _____ 7. The pH of a neutral solution is _____ 7. [3]

2. a. Copy and complete the table to show whether each mixture is acidic, alkaline, or neutral.

Mixture	pH	Acidic, alkaline, or neutral?
orange juice	3	
milk	7	
cola drink	2	
sweat	5	
indigestion medicine	9	

   [5]

   b. Name one substance in the table above that could be used to neutralise orange juice. [1]

   c. Name one substance in the table that could be used to neutralise the indigestion medicine. [1]

3. This question is about litmus indicator.

   a. Copy and complete the table below.

Type of solution	Colour of litmus indicator
acidic	
alkaline	

   [2]

   b. A student has 100 cm³ of an acid. She adds a few drops of litmus indicator. The student then adds 500 cm³ of an alkaline solution of the same concentration as the acid. Predict the colour changes she would observe. [2]

4. Choose one word from the list below to answer each question part.

   carbon dioxide
   hydrogen
   oxygen

   a. This gas makes a lighted splint go out with a squeaky pop. [1]

   b. This gas relights a glowing splint. [1]

   c. This gas makes limewater milky. [1]

5. Read the information in the box. Then answer the questions below.

   > Your blood is slightly alkaline. Its pH is always 7.4. You would be ill if your blood pH was higher than 7.6 or lower than 7.2.
   >
   > The pH of your urine changes to help adjust blood pH:
   >
   > - If your blood gets too acidic, extra acid comes out in your urine. Your urine pH gets lower.
   >
   > - If your blood gets too alkaline, extra alkali comes out in your urine. Your urine pH gets higher.

   a. Explain why your urine pH gets lower when your blood is too acidic. [1]

   b. Explain what happens to your urine pH if your blood is too alkaline. [1]

   c. A hospital patient has a blood pH of 7.5. Explain how her body tries to get the blood pH back to normal. [1]

6. The table gives the preferred soil pH of some plants.

Plant	Preferred soil pH
pineapple	4.5 to 5.5
banana	5.5 to 6.5
sugar cane	5.5 to 6.5
maize	5.5 to 7.0
cassava	4.5 to 7.5

a. Name the one crop in the table that can grow well in a slightly alkaline soil. [1]

b. A farmer tests the soil pH on her farm. Its pH is 5.0. Suggest two crops she could try growing. [2]

c. A farmer knows her soil is acidic, but does not know the soil pH. Suggest one crop she could try growing. [1]

d. The soil pH on another farm is 7.0.

   i. Suggest two crops that might grow well on this soil. [2]

   ii. The farmer on this farm wants to grow bananas. Name the type of substance that he should add to the soil. [1]

7. A student wants to make an indicator from red cabbage. The apparatus available to the student is shown below.

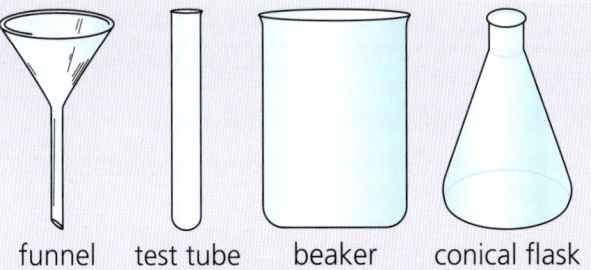

funnel   test tube   beaker   conical flask

a. The student first heats a mixture of pure water and chopped cabbage. Name the best apparatus in which to do this. [1]

b. Next the student filters the mixture. She wants to keep the solution. Name the best two pieces of apparatus for this stage. [2]

c. The student adds her indicator to small amounts of acid and alkali to observe its colours in the two solutions. Name the best apparatus in which to do this. [1]

8. Saima eats a sugary sweet. The pH of the saliva in her mouth changes. The graph shows how the pH changed.

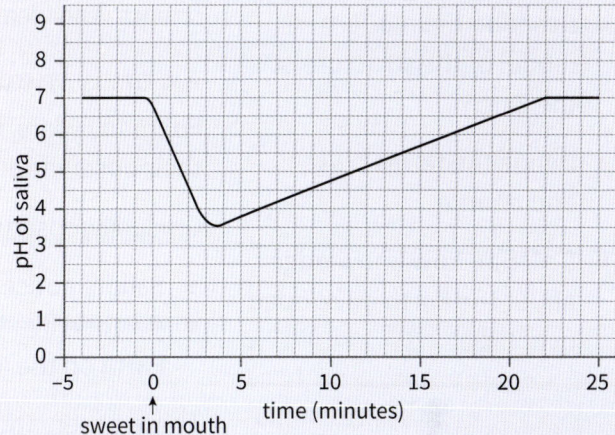

a. i. Give the pH of Saima's saliva before she ate the sweet. [1]

   ii. Give the pH of Saima's saliva 3 minutes after eating the sweet. [1]

   iii. Give the time at which Saima's saliva returns to its original pH. [1]

d. Copy and complete the sentence below. When eating the sweets, Saima's saliva became _____ acidic. [1]

e. Teeth are damaged when saliva pH is 5.5 or below. Give the number of minutes that Saima's teeth are at risk of damage. [1]

f. After her meal, Saima used toothpaste to clean her teeth. The toothpaste neutralised her saliva. State whether the toothpaste is acidic, neutral, or alkaline. [1]

Science in context 6.1

Models of the Earth

Imagine you could dig a hole more than 6000 km deep, to the centre of the Earth. What would you find?

For many years, scientists have been curious about the structure of the Earth. They made observations and collected data. They thought carefully about them. They created models to explain their observations. As you know, a model in science is an idea that explains observations and helps in making predictions.

Objectives

- Describe the modern model of the structure of the Earth
- Define the terms *crust* and *mantle*
- Describe how knowledge about the Earth developed over time

An early model of the Earth

For many years, people thought the Earth was flat. They based this idea on their observations.

Gradually, observations made people think that the flat Earth model was wrong:

- Sailors noticed that ships appeared to sink as they go over the horizon.
- Aristotle lived more than 2000 years ago. He saw that the shadow of the Earth on the Moon is round.

These observations led to a new model of the Earth, as a sphere.

▲ *Ships appear to sink as they go over the horizon.*

▲ *Observations from space give further evidence that the Earth is a sphere.*

The modern model of the structure of the Earth

Scientists used observations and data to create the modern model of the structure of the Earth.

The model states that the Earth is made up of four layers:

- A solid **crust**, made of different types of rock. The crust does not flow.
- The **mantle**, which is solid but flows slowly.
- The liquid **outer core**, which is mainly iron and nickel.
- The solid **inner core**, also mainly iron and nickel.

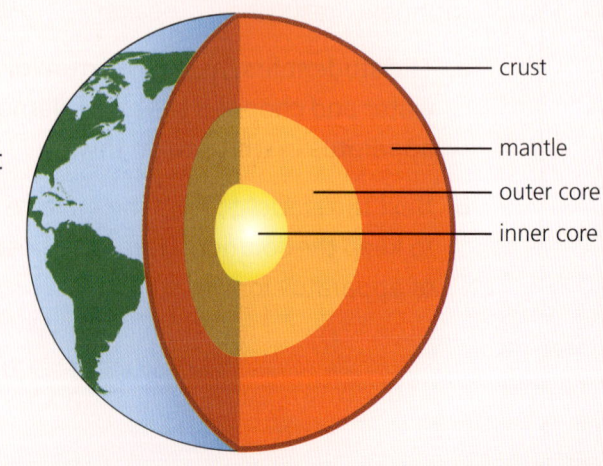

How do we know about the structure of the Earth?

Evidence from rocks

Over many years, scientists studied rocks on the surface of the Earth. They observed rocks under oceans. They examined rocks brought to the surface by volcanoes. Their observations helped to create the modern model of the Earth.

Evidence from earthquakes

Shock waves from earthquakes also provided evidence. In 1909, there was an earthquake in Croatia. Andrija Mohorovičić studied shock wave patterns at different places. He thought about the patterns. He realised that the shock waves had come through two different layers. He had found the crust and the mantle.

In the 1930s, Inge Lehmann also examined shock wave patterns. She could not explain them using the model of the time – that the Earth's core is the same all the way through. Instead, she created a new model – that the core has two parts. Inge Lehmann had discovered the solid inner core. Her finding is an important contribution to the modern model of the Earth.

▲ *Inge Lehmann discovered the inner core.*

Evidence from mines

Temperatures increase from crust to core. The core, 6000 km below the surface, may be hotter than the surface of the Sun.

Evidence for the temperature increase comes partly from mining engineers. Some of the deepest mines in the world are in South Africa. They are nearly 4 km deep. Here, the rock temperature is about 60 °C. The air is cooled so that miners can work.

▲ *Deep mines are hot.*

Questions

1. Name the layers of the Earth, starting with the inner core.
2. Compare the crust to the mantle.
3. Describe two pieces of evidence that suggest that the Earth is a sphere.
4. Describe two types of evidence that scientists used to help to develop the modern model of the structure of the Earth.

Key points

- The Earth consists of the crust, mantle, outer core, and inner core.
- Observations from many scientists contributed to the modern model of the structure of the Earth.

6.2 Plate tectonics

What does an extinct reptile tell us about the Earth as it was, thousands of years ago?

Objectives

- Define the terms *tectonic plate* and *peer review*
- Describe the model of plate tectonics
- Describe how building on others' work helps scientists to develop understanding

Moving continents

A hypothesis

In the early 1900s, Alfred Wegener saw that Africa and South America look as if they fit together. He wondered if they had once been joined together, and are now moving apart. He looked for evidence to support his idea.

▲ *A* Mesosaurus

Wegener found different types of evidence:

- He read articles by other scientists about rock types in Africa and South America. There are similar rocks at the edges of the continents, where they might have been joined.
- He heard that scientists had found fossils of the same reptile in Africa and South America. The reptile, *Mesosaurus*, lived in shallow water. It could not swim across the ocean.

▲ The shapes of South America and Africa

▲ There are Mesosaurus *fossils in Africa and South America.*

By 1912, Wegener had enough evidence to make a hypothesis. He said that the continents were once a single piece of land. The land broke up. The pieces drifted apart, making continents.

Rejection

At first, other scientists rejected Wegener's hypothesis. They did not know how continents could drift apart. They did not trust Wegener because he mainly studied the weather, not rocks.

Plate tectonics

Over time, different scientists found more evidence to support Wegener's idea. They developed it further. By 1964, many scientists accepted the **model of plate tectonics**.

The model of plate tectonics says that the Earth's crust and the uppermost mantle are made of about 12 huge slabs of solid rock. These are **tectonic plates**. The plates rest on top of the lower part of the mantle.

The lower mantle is heated from deep inside the Earth. The heat makes the lower mantle flow, even though it is solid. The solid tectonic plates move with the flowing mantle below. They move slowly. Their speed is a few centimetres per year.

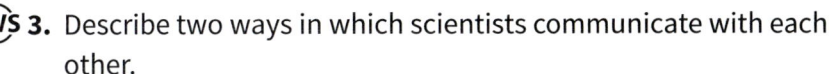
▲ *Map of tectonic plates*

Science in context

How do scientists build on each other's ideas?

Wegener used evidence from other scientists to help make his hypothesis. Other scientists built on Wegener's ideas to create the model of plate tectonics.

Most scientists do not work in isolation. They build on the work of scientists before them. They collaborate (work jointly) with others, both nearby and far away.

▲ *Scientists at a conference*

Scientists share and discuss their evidence and ideas in different ways:

- By talking to and messaging each other.
- By giving talks at conferences or webinars.
- By showing posters at conferences.
- By writing papers for scientific journals. Other expert scientists check the papers. This is **peer review**. If a paper is good enough, it is published in a scientific journal. The journals are published online and some are also published in print.

Key points

- A tectonic plate is one of about 12 huge slabs of rock on the Earth's surface. It is made of crust and the top of the mantle.
- The model of plate tectonics says that tectonic plates rest on – and move with – the lower mantle.
- Scientists build on and review each other's work.

Questions

1. Write definitions for *tectonic plate* and *peer review*.
2. Describe two pieces of evidence suggesting that Africa and South America were once joined together.
3. Describe two ways in which scientists communicate with each other.

6.3 The restless Earth

Objectives

- Describe how earthquakes happen at plate boundaries
- Describe how fold mountains are made

30 October, 2020: A magnitude 7.0 earthquake strikes Turkey. At least 116 people die. More than 1000 people are injured. By the end of the year, there had been 1400 aftershocks.

What caused the disaster? Could it have been predicted?

▲ After the 2020 earthquake in Turkey.

Thinking and working scientifically

Using the model of plate tectonics

The model of plate tectonics explains observations. For example, the model explains:

- why earthquakes happen
- how mountains form.

Earthquakes

What makes earthquakes?

Earthquakes happen when tectonic plates move against each other suddenly.

At the San Andreas Fault in California, USA, two tectonic plates are sliding past each other in opposite directions.

Because of friction, the plates cannot slide smoothly. Sometimes they get stuck. Huge forces build up. Eventually, the two plates overcome the frictional forces. The plates slip suddenly. There is an earthquake.

The place where the plates slip is the focus of an earthquake. Shock waves spread from there. The shock waves make buildings collapse. Collapsing buildings may kill or injure people.

Can we predict earthquakes?

Earthquakes are common at moving plate boundaries. No one can predict when plates will suddenly slip. If you live near a plate boundary, you can expect an earthquake at any time.

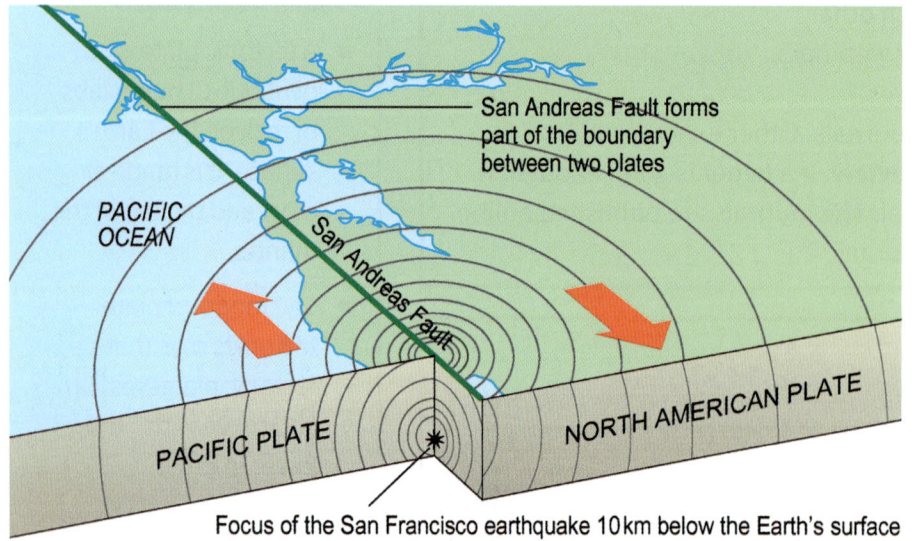

▲ The focus of an earthquake on the San Andreas Fault.

Models of the Earth

There are some clues that an earthquake may happen soon:

- There may be more – or bigger – vibrations in the Earth's crust. Seismometers pick these up.
- Large amounts of radon gas may suddenly escape from cracks in the crust.

Making mountains

Fold mountains may form where tectonic plates push together.

Sixty million years ago, India was not joined to Asia. The Indian plate was moving north, at a speed of 9 metres per century. The Eurasian plate was moving south. Between 40 and 50 million years ago, India and Asia collided. The edges of the continents crumpled together and piled up. The jagged peaks of the Himalayas were forming.

Today, Mount Everest in the Himalayas has risen to a height of 9 km. The Himalayas continue to get taller as the two tectonic plates push together.

We can use the model of plate tectonics to predict that the Himalayas will get taller in future as the plates push together.

▲ A seismometer shows bigger vibrations when there is an earthquake.

▲ Some jagged peaks in the Himalayas

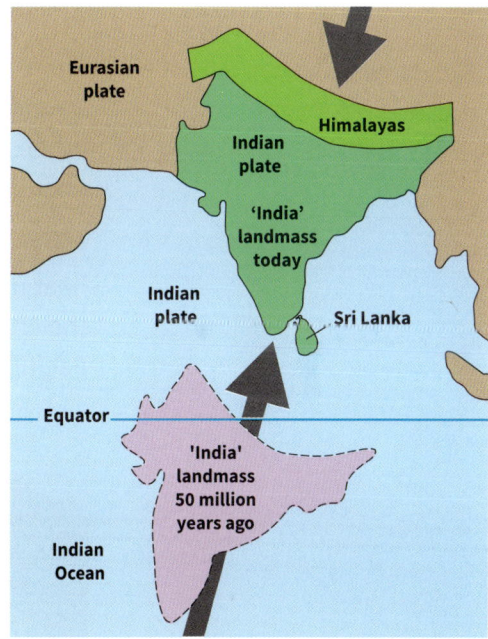
▲ Plate tectonics shows that the plates are pushing together.

Key points

- Earthquakes happen when tectonic plates move against each other suddenly.
- Fold mountains may form when tectonic plates push together.

Questions

1. Describe how an earthquake may happen.
2. The Himalayas are fold mountains. Describe how they were formed.

6.4 Volcanoes

Objectives

- Define the terms *volcano* and *lava*
- Explain why volcanoes form near plate boundaries
- Describe how volcano scientists collaborate

A volcano flings ash into the sky. Hot rocks and poisonous gases tumble out. Bubbling lava flows quickly down its slopes. What is a volcano? And why did this one erupt when it did?

What are volcanoes?

A **volcano** is an opening in the Earth's crust. Liquid rock and other materials escape from it.

The rock under a volcano is hot. The rock melts. Under the ground, liquid rock is called magma. Magma collects under and inside the volcano. When the volcano erupts, liquid rock escapes from it. Above ground, liquid rock is called **lava**.

An erupting volcano also flings out solid materials (like volcanic bombs and ash) and gases (like sulfur dioxide, carbon dioxide, and water vapour).

Where are volcanoes?

There are about 500 active volcanoes in the world. Most volcanoes are at plate boundaries, where tectonic plates are moving towards – or away from – each other. They form at plate boundaries because underground rock can get hot enough to melt here.

A few volcanoes are not at plate boundaries. They are at hot spots, where heat melts rock in the mantle. The liquid rock pushes through cracks in the crust, forming volcanoes. The volcanoes in Hawaii, in the middle of the Pacific Plate, are hot spot volcanoes.

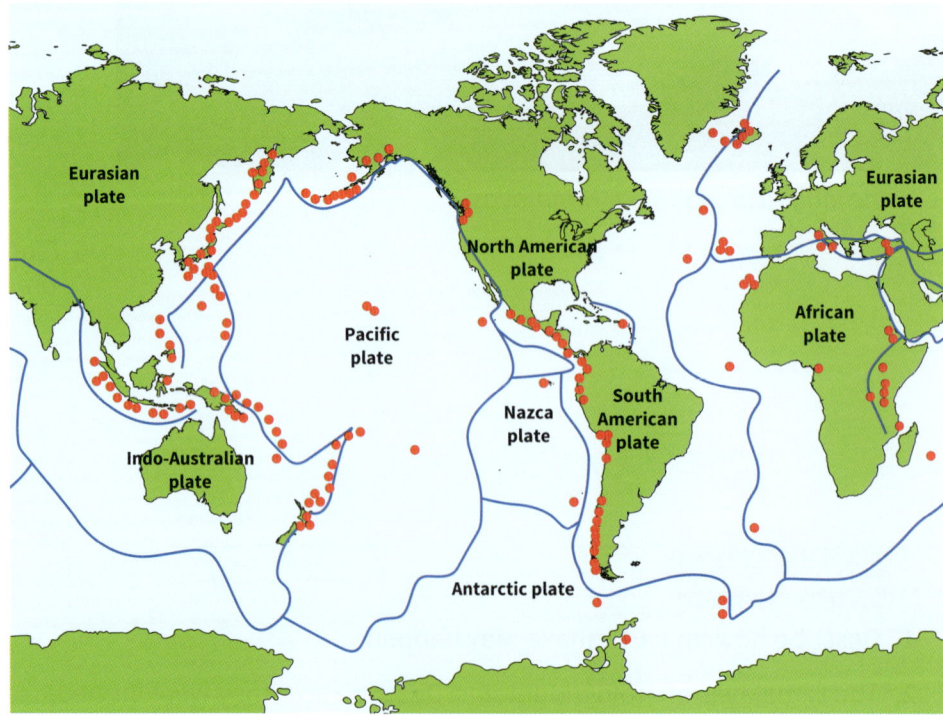

▸ Most volcanoes are at plate boundaries.

108

Models of the Earth

🏠 Science in context

How do scientists collaborate to predict eruptions?

It is not possible to know exactly when – or how – a volcano will erupt. But it is possible to make good predictions.

Vulcanologists study volcanoes. They make observations and measurements. They look for patterns in their data. They use these patterns to help make predictions.

Different vulcanologists are expert in different things. They **collaborate** (work together) to predict what a volcano will do next. They may tell people when to leave their homes to escape an eruption.

My tilt meter measures the steepness of the volcano slope. I predict that, when the steepness changes, the volcano is more likely to erupt.

I measure the amount of sulfur dioxide gas coming out of the volcano. I predict that, if the amount of sulfur dioxide increases suddenly, the volcano will erupt.

I monitor earth movements near the volcano. I predict that, if earth movements increase, the volcano will soon erupt.

Explanation:
Changes in the shape of a volcano show that magma is moving inside the volcano. Moving magma means an eruption is likely.

Explanation:
Magma contains dissolved sulfur dioxide gas. The gas escapes when magma rises to the surface. Extra sulfur dioxide shows that magma is near the surface. An eruption is likely.

Explanation:
Earth movements may be caused by magma pushing up against the surface rock. If magma is moving, an eruption is likely.

🔍 Questions

1. Name six substances that come out of volcanoes.
2. Explain why most volcanoes form near tectonic plate boundaries.
3. List three things that vulcanologists may monitor at volcanoes.
4. Suggest why vulcanologists collaborate.

📖 Key points

- A volcano is an opening in the Earth's crust that liquid rock and other materials escape from.
- Most volcanoes are near plate boundaries.
- Vulcanologists collaborate to predict when a volcano will erupt.

Review 6.5

1. The diagram shows the structure of the Earth.

   a. Copy the diagram and add labels. [4]

   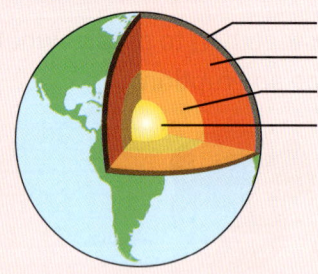

   b. Copy and complete the table below. [4]

Layer	Solid, liquid, or gas?
crust	
inner core	
mantle	
outer core	

   c. Name the part of the Earth that is solid, but flows slowly. [1]

2. The pie chart shows the percentage by mass of the different elements that make up the substances in the Earth's crust.

   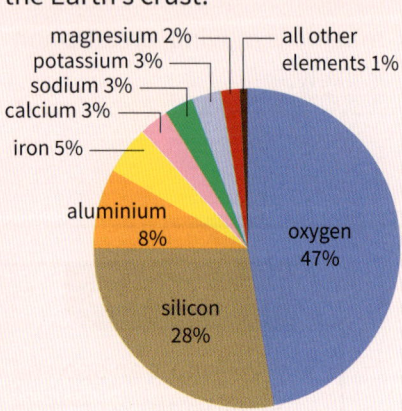

   a. Write down the names and chemical symbols of six metal elements in the Earth's crust. [6]

   b. Name the metal element in the Earth's crust that is present in the greatest amount. [1]

   c. Name the two elements that together make up 75% of the mass of the Earth's crust. [1]

3. The map shows the Earth's tectonic plates.

   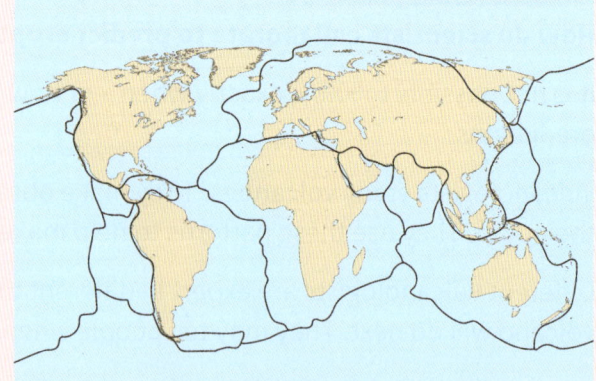

   Copy and complete the sentences. Choose words from the list below. Use each word once, more than once, or not at all. [6]

twelve	flows	solid	mantle
metres	centimetres		liquid

   There are about _____ tectonic plates. A tectonic plate is made of _____ rock. It consists of the Earth's crust and uppermost _____. The tectonic plates rest on the lower _____, which _____. The tectonic plates move with the flowing mantle below. Their speed is a few _____ each year.

4. This question is about volcanoes.

   a. Write the definition for a volcano. [1]

   b. Name one liquid material that comes out of volcanoes. [1]

   c. Name three gases that come out of volcanoes. [3]

   d. Explain why volcanoes are often found at plate boundaries. [1]

5. A hard-boiled egg can help us to imagine the scientific model of the Earth.

   a. State which parts of the egg represent the crust, mantle, and core. [3]

   b. Suggest strengths and limitations of the egg model. [3]

   ◀ A hard-boiled egg

6. The map shows two continents - South America (A) and Africa (B).

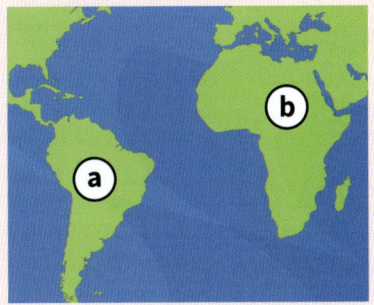

   a. Explain how the map provides evidence that South America and Africa were once joined together. [2]

   b. List two more pieces of evidence that South America and Africa were once joined. [2]

7. The road in the picture was damaged in an earthquake.

   The steps below explain why an earthquake happens. List the letters of the steps in the correct order. [5]

   A Because of friction, the plates get stuck.
   B Two tectonic plates slide past each other.
   C The plates slip suddenly.
   D Shock waves spread from the earthquake.
   E There is an earthquake.
   F Huge forces build up.

8. The map shows the tectonic plates around South America, as well as some cities in South America.

   a. Explain why there is a greater risk of earthquakes in Santiago than in Buenos Aires. [2]

   b. Name one other city on the map that is at high risk of earthquakes. [1]

9. Look at maps A and B.

▲ **Map A** shows the Atlas Mountains in orange. The mountains are in North Africa. They are not really red.

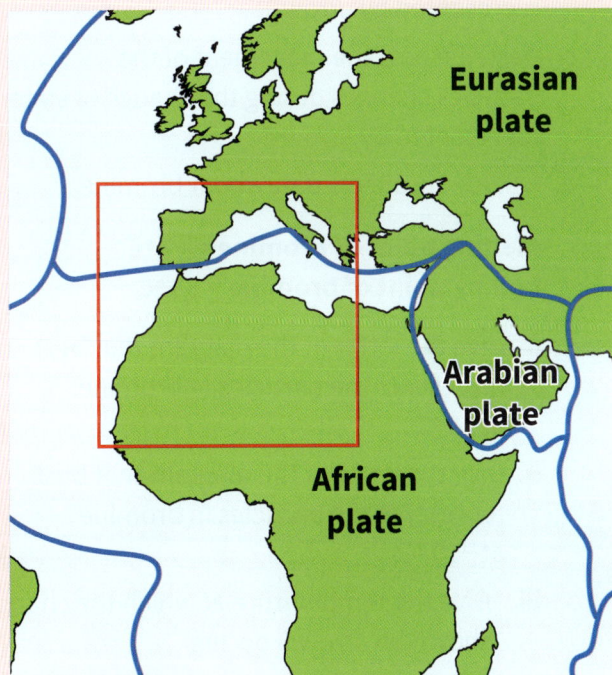

▲ **Map B** Shows some tectonic plates. The area in the red box shows the area of map A.

   a. The Atlas Mountains are fold mountains. Name the two tectonic plates that pushed together to form the Atlas Mountains. [2]

   b. A student predicts that the Atlas Mountains will continue to get taller. Suggest an explanation for her prediction. [1]

Stage 7 Review

1. Bromine can exist as a solid, liquid, or gas. The diagrams show the arrangements of particles in each of these states.

   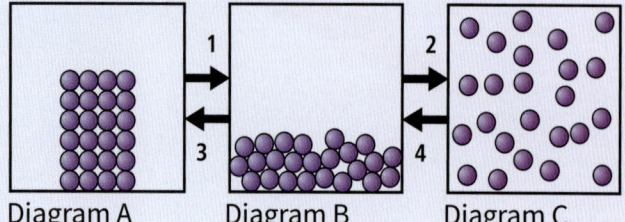

   Diagram A   Diagram B   Diagram C

   a. Give the name of the state represented by diagram A. [1]
   b. Give the name of the change of state that is represented by arrow 1. [1]
   c. Describe how the movement of the particles changes during the change of state represented by arrow 4. [2]
   d. A student collects data about bromine.

      melting point of bromine = −7 °C
      boiling point of bromine = 59 °C

      i. Give the letter of the diagram that best represents the particles in bromine at 20 °C. [1]
      ii. Give the letter of the diagram that best represents the particles in bromine at −10 °C. [1]
      iii. Give the temperature at which the change shown by arrow 3 occurs. [1]

2. The table gives the melting and boiling points of some elements.

Element	Melting point (°C)	Boiling point (°C)
argon	−189	−186
bromine	−7	59
calcium	850	1487
gallium	30	2400
zirconium	1850	3580
technetium	2200	3500

   a. Name the elements that are in the solid state at 20 °C. [1]
   b. Name the elements that are in the liquid state at 40 °C. [1]
   c. Give the state of calcium at 800 °C. [1]

3. Use the words and phrases to complete the sentences below. You may use each word or phrase once, more than once, or not at all.

vibrate on the spot	gas
far apart	liquid
close together	solid
a little	much
move around from place to place	
move around, sliding over each other	

   Copper exists in three states, solid, liquid, and _____ In the solid state, its particles _____ The particles are _____ When copper melts, it changes state from _____ to _____ Its particles start to _____ . They get _____ further apart.

   If copper is heated to 1084 °C it changes from the liquid to the _____ state. Its particles get _____ further apart, and they start to _____ [10]

4. Give the chemical symbols of the elements in the list. [6]

   a. carbon
   b. silicon
   c. germanium
   d. arsenic
   e. cadmium
   f. silver

5. Give the names of the elements with these chemical symbols. [4]

   a. F
   b. Fr
   c. Fe

112

6. Match the names of the substances to their formulae.

Name	Formula
carbon dioxide	He
copper sulfate	CO_2
carbon monoxide	N_2
nitrogen	$CuSO_4$
helium	CO

[5]

7. From the list below, write the **five** properties that are typical of metals.

high melting point	shiny
dull appearance	malleable
poor conductor of heat	sonorous
low boiling point	brittle
good conductor of electricity	

[5]

8. The bar chart shows the melting points of some elements. Each element is represented by a letter. The letter is not the chemical symbol of the element.

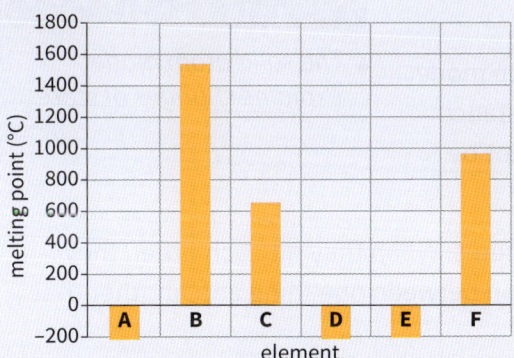

a. Give the letter of the element on the bar chart with the highest melting point. [1]

b. Give the letters of three elements on the bar chart that are most likely to be metals. [1]

c. Give the letters of three elements on the bar chart that are least likely to conduct electricity. [1]

9. This question is about the pH of the water in three East African lakes.

The pH range of the water in each lake is given in the table below.

Name of lake	pH range of lake water
Lake Malawi	7.8–8.6
Lake Tanganyika	7.2–8.6
Lake Victoria	8.6–9.5

a. Give the name of the lake with the most alkaline water. [1]

b. A student collected a sample of water from Lake Victoria. She added a few drops of universal indicator to the water. Use data from the table, and the diagram below, to predict the colour of this mixture. [1]

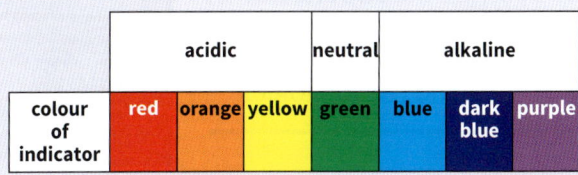

c. Read the paragraph in the box. It is from a book about Lake Malawi.

> The pH of Lake Malawi water is different in different parts of the lake. In calm bays, more carbon dioxide dissolves in the water. Carbon dioxide gas is acidic, so the water in calm bays is more acidic.
>
> Where the water is not calm, the water is more alkaline. This is because less carbon dioxide is dissolved in the water.

Use the information in the box, and data from the table, to predict the pH in a calm bay of Lake Malawi. [1]

d. Predict how the pH of the lakes might change if the rain in the region became more acidic. [1]

e. The picture shows one species of tilapia fish.

Tilapia prefer to live in water of pH 6 to 9. However, many can survive in water of pH 5 to 10. Predict which of the lakes in the table are suitable for tilapia. [1]

7.1 Inside atoms

Have you ever watched – or played – snooker, pool, or billiards? The balls are solid spheres.

Objectives

- Describe the structure of an atom
- Define and use analogies

Thinking and working scientifically

Atoms and analogies

Until the early 1900s, scientists modelled atoms as solid spheres. As you know, a model is an idea that explains observations and helps in making predictions.

The solid atom model is good enough to explain the different properties of a substance in the solid, liquid, and gas states. The model is good enough to make predictions. For example, when a substance changes state, how will its properties change? The model cannot be used to explain or predict chemical reactions.

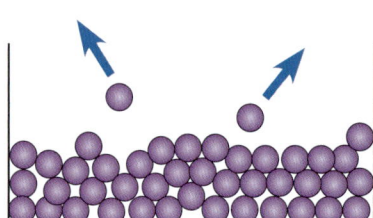

▲ The solid atom model explains changes of state.

A teacher says that atoms are like pool balls. They are solid spheres. When they hit each other, they change direction. This is an analogy. An **analogy** is a comparison between one thing and another that helps to explain something. It can be used as a model.

Analogies are useful in science. They help to explain things that are very small, or very big. In Review 6.5, the egg is an analogy for the structure of the Earth.

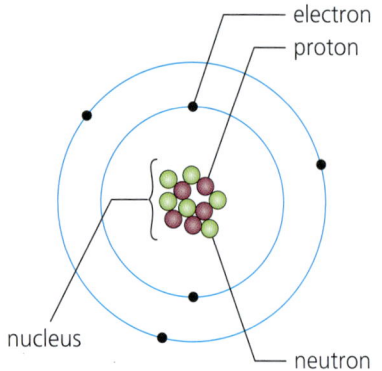

▲ Atoms are made up of three types of sub-atomic particles. Not to scale.

In the early 1900s, scientists wanted to find out what is inside an atom. They collected evidence and thought about it carefully. Units 7.2 and 7.3 describe what they did.

By 1932, scientists had developed a new model. This is the model you will use. You have already seen it in physics. The model explains how atoms join together. It also explains chemical reactions.

The model states that atoms are made up of tiny **sub-atomic particles**. There are three types of sub-atomic particle – **protons**, **neutrons**, and **electrons**.

Protons and neutrons make up the nucleus. The **nucleus** is at the centre of an atom. Electrons orbit outside the nucleus.

Mass and charge

The table gives the relative mass and charge of protons, neutrons, and electrons.

Sub-atomic particle	Relative mass	Relative charge
proton	1	+1
neutron	1	0
electron	$\frac{1}{2000}$	−1

The table compares the masses of protons, neutrons, and electrons. Their actual masses are tiny. The mass of a proton is about 0.000 000 000 000 000 000 000 000 0017 kg.

Thinking and working scientifically

Another analogy

Nearly all the mass of an atom is in its nucleus. But a nucleus is tiny compared to its atom. An analogy is helpful here. If you imagine an atom to be the size of a football stadium, the nucleus is the size of a pea at its centre.

▲ *If an atom was the size of a football stadium, the nucleus would be the size of a pea.*

Atoms and electrical charge

Why do atoms have no electric charge?

Atoms contain charged sub-atomic particles. But they have no net charge. They are neutral. This is because, in an atom, the number of protons is equal to the number of electrons.

For example, a boron atom (shown opposite) has:

- 5 positive protons
- 5 negative electrons
- 6 neutral neutrons.

What holds an atom together?

The nucleus of an atom is positively charged. The electrons are negatively charged. An individual atom is held together by the electrostatic attraction between its positively charged nucleus and its negatively charged electrons.

Key points

- Atoms are made up of protons, neutrons, and electrons.
- Protons and neutrons make up the nucleus. Electrons orbit outside.
- An analogy is a comparison that helps to explain something.

Questions

1. Give the relative charge and mass on: a proton, a neutron, and an electron.
2. Name the type of force that holds an atom together.
3. Describe an analogy that helps to explain the size of a nucleus compared to its atom.
4. Draw a beryllium atom. It is made up of four protons, four electrons, and five neutrons.
5. Explain why a beryllium atom is electrically neutral.

Thinking and working scientifically 7.2

Objectives

- Describe the steps that Thomson took to develop his 1904 model of the atom
- Evaluate the plum pudding analogy for Thomson's atomic model

Discovering electrons

No one can see inside atoms. So how did scientists discover sub-atomic particles? How did they develop models of the atom?

Finding electrons

Cathode rays

In the 1800s, scientists investigated gases. They took sealed tubes containing tiny amounts of gas. They set up an electric circuit and applied huge voltages.

Amazingly, the gases conducted electricity. One day, Johann Hittorf noticed a green glow on the screen. The glow, he said, was caused by rays from the negative electrode. The rays travelled through the gas and hit the screen. These were cathode rays.

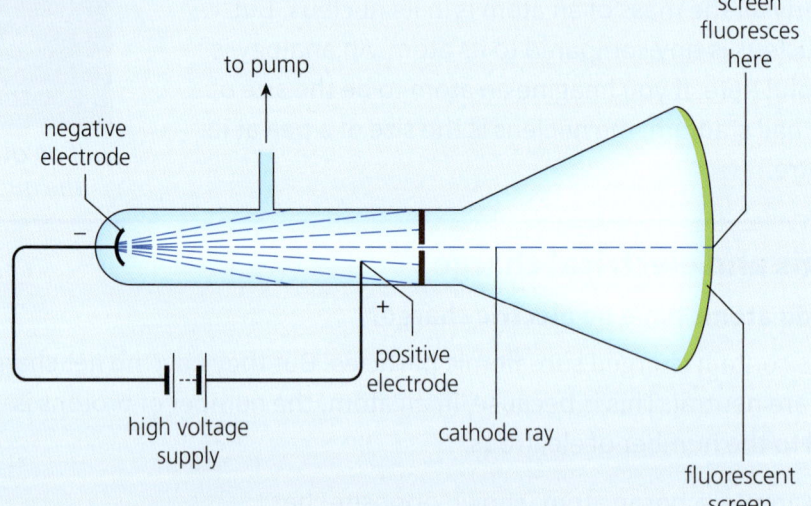

▲ When the voltage is high, cathode rays travel through the gas.

Planning an investigation

Scientist Joseph John Thomson thought about cathode rays. He asked a scientific question:

Scientific question: What **are** cathode rays?

Thomson thought about his question. He came up with a hypothesis. As you know, a hypothesis is a possible explanation. It is based on evidence, and can be tested further.

Hypothesis: Cathode rays move towards a positive electrode, so they might have a negative electrical charge.

Thomson planned an investigation. He made a prediction:

Prediction: If I pass cathode rays between electrically charged pieces of metal, the rays will change direction. The rays will bend towards the positively charged metal.

Carrying out the investigation

Thomson carried out his investigation. He made observations. The rays bent towards the positively charged metal. His prediction was correct.

Analysing the evidence

Thomson thought about his observation. It supported his hypothesis. He made a conclusion.

Conclusion: Cathode rays are charged. Their charge is negative.

Thomson asked more questions. He did some more thinking. He did some different investigations, and collected more evidence.

He concluded that cathode rays are made up of particles. All the particles have the same – tiny – mass. They all have the same electrical charge. By 1897, Thomson had discovered the first sub-atomic particle – the electron.

A new model of the atom

Thomson knew that electrons are negatively charged. He also knew that atoms have no overall electrical charge.

In 1904, Thomson used this evidence to suggest a new model of the atom. An atom is a positively charged sphere, he said. There are negative electrons embedded in the sphere.

The model explains the observations in cathode ray experiments.

Plum pudding analogy

Some people said that Thomson's model reminded them of a plum pudding. The plum pudding is an analogy for Thomson's model.

▲ Thomson's model of the atom

▲ A plum pudding

Like all analogies, the plum pudding analogy is not perfect:

- Electrons are negatively charged. But the plums in the pudding have no charge.
- In Thomson's model, the atom is a positively charged sphere. But the pudding has no charge.

The story continues

The story of the development of atomic models does not end here. See Unit 7.3 to find out more.

Questions

1. Thomson suggested that cathode rays are electrically charged. Describe the evidence that supports this prediction.
2. List the steps that Thomson took in his investigation that led to the development of his model.
3. The plum pudding analogy helps to explain Thomson's model. Describe two strengths and two weaknesses of the plum pudding analogy.

Key points

- Thomson observed negatively charged rays in gases. After this, he developed a new model of the atom.
- The plum pudding analogy for Thomson's model of the atom has strengths and weaknesses.

Science in context 7.3

Objective
- Describe how scientists build on others' work to develop scientific knowledge over time

Finding the nucleus

Thomson told other scientists about his plum pudding model. The other scientists thought about it. There was good evidence for negatively charged electrons. But what about their arrangement in a positively charged sphere? Was there evidence for this?

Testing the plum pudding model

Ernest Rutherford lived in New Zealand until he was 23. In 1895, he moved to England to study under Thomson.

▲ *In the plum pudding model of an atom, electrons are embedded in a sphere.*

Rutherford planned an investigation to test Thomson's plum pudding model. He made a prediction.

Prediction: We will fire positively charged particles at a piece of gold foil. If the plum pudding model is correct, most of the positive particles will go straight through the foil. A few will pass close to negative electrons. These positive particles will change direction slightly.

Rutherford did not work alone. He collaborated with Hans Geiger and Ernest Marsden. They set up the apparatus below.

▲ *Gold foil*

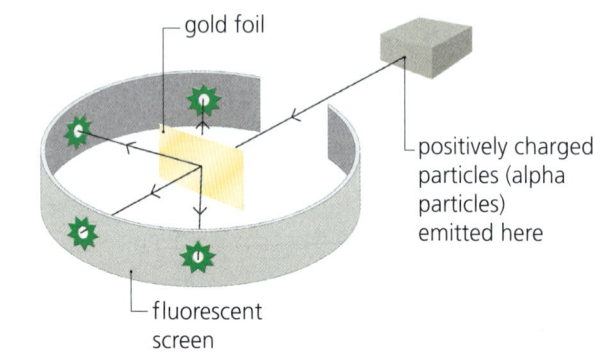

▲ *Geiger and Marsden used this apparatus to test the plum pudding model in 1909.*

The scientists fired positive particles at the gold. The results were amazing. Around one in every 10 000 particles bounced backwards off the foil. Rutherford wrote:

It was quite the most incredible event that has ever happened to me in my life. It was almost as incredible as if you fired an artillery shell [a weapon] at a piece of tissue paper and it came back and hit you.

The scientists discussed their observations. Rutherford's prediction was wrong. He had based his prediction on Thomson's plum pudding model. This, said Rutherford, means the plum pudding model must be wrong.

Rutherford thought about the problem. Could he come up with a better model to explain the evidence?

A new model for the atom

By 1911, Rutherford had developed a new model:

- Atoms have a central nucleus. Most of the mass of an atom is in its nucleus. The nucleus is positively charged.
- A nucleus is surrounded by empty space in which electrons move.

Rutherford's new model explained the surprising observations:

- The positive particle that bounced back had hit a nucleus.
- The positive particles that went straight through the foil had passed through the empty space between nuclei.

Rutherford wrote a paper about his new model. He also described the problems with the plum pudding model. The paper was published in a scientific journal.

Other scientists read about Rutherford's model. At first, they did not accept it. Over time, scientists like Niels Bohr developed the model further. Bohr discovered that electrons move in orbits.

Inside the nucleus

Scientists wondered about the nucleus. Was it made up of smaller particles?

Rutherford collected evidence. He fired positive particles into the air. Tiny, new positively charged particles were made. Where did they come from? Rutherford realised that the new particles came from the nuclei of nitrogen atoms. The tiny particles were protons.

The next year, Rutherford suggested that protons were not the only particles in the nucleus. Did nuclei also contain particles with mass, but no charge?

Scientist James Chadwick investigated this question. By 1932, he had an answer. A nucleus contains neutrons as well as protons.

The model for an atom now included:

- positive protons and neutral neutrons in the nucleus
- negative electrons orbiting outside the nucleus.

This is the model you use today. Several scientists built on each other's work to develop the model.

▲ *Marie Curie and Ernest Rutherford organised a conference in 1913 for scientists to discuss models for atoms.*

Questions

1. **a.** Name six scientists who built on each other's work to develop the model of the atom you use today.

   b. Suggest why one scientist alone could not have developed this model.

2. Describe Rutherford, Geiger, and Marsden's evidence for the nucleus.

Key points

- Scientists built on each other's work to develop models of the atom.

Science in context 7.4

Inside sub-atomic particles

By 1932, scientists knew that most of the mass of an atom is in its nucleus. The nucleus contains protons and neutrons. Electrons move around an atom, outside its nucleus.

Scientists continued to ask questions. What makes up sub-atomic particles, such as protons and electrons? Are they made up of even smaller particles?

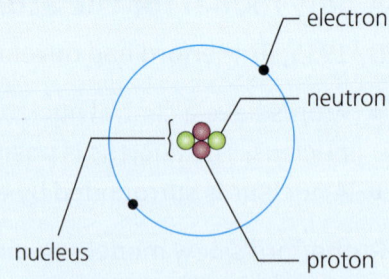

▲ *A helium atom has two protons and two neutrons in its nucleus. Two electrons move around the atom, outside its nucleus.*

Collaborating to find answers

Scientists suggested answers to these questions. They used maths to support their ideas. Satyendra Nath Bose thought about sub-atomic particles and energy. He wrote to Albert Einstein about his ideas. Together, Bose and Einstein predicted a new state of matter. They called it the Bose–Einstein condensate. The new state would only exist at very low temperatures.

Other scientists built on the ideas of Bose and Einstein. In 1964, six scientists, including Peter Higgs, made a hypothesis. There is a particle, they said, that gives sub-atomic particles their mass. They called the particle the Higgs boson.

Testing the hypothesis

Scientists began looking for evidence to support the hypothesis that Higgs bosons exist.

Thousands of scientists from more than 100 countries collaborated. Different scientists were expert in different things. But they had one purpose – to design and build apparatus to learn more about sub-atomic particles. The result of their work – the Large Hadron Collider (LHC) – was ready to use by 2010.

The LHC is a 27 km circular tunnel, deep underground. It makes beams of high-energy protons. Inside the tunnel, huge magnets guide the protons around the circle, in both directions.

Objectives

- Describe how scientists build on others' work to develop scientific knowledge over time
- Explain why and how scientists collaborate

▲ *Satyendra Nath Bose was born in Kolkata, India, in 1894.*

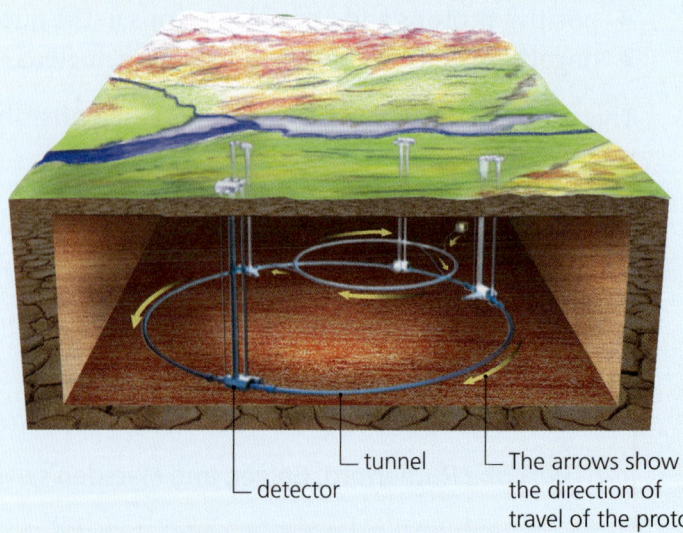

▲ *The Large Hadron Collider*

Inside atoms

The protons travel faster and faster. They collide with each other, and break up. Four huge detectors follow the collisions. They detect the particles made in the collisions.

The detectors collect data. They send the data to hundreds of scientists in many countries. The scientists are expert in different things.

The scientists study the evidence. They ask questions, such as: Does the evidence support the hypothesis that Higgs bosons exist?

The scientists may collaborate to make conclusions. They write papers about their work. Other expert scientists check the papers. They may suggest corrections and improvements. This is peer review. If a paper is good enough, it is published in a scientific journal. Other scientists read the paper. The paper may give them ideas for new investigations.

▲ The picture represents two protons colliding.

Evidence for the Higgs boson

On 4 July 2012, LHC scientists made an announcement. They had detected a new boson. The new boson behaves as they had predicted a Higgs boson would.

The evidence supports the hypothesis made nearly 50 years earlier. There is a particle that gives sub-atomic particles their mass. This particle is the Higgs boson.

What next for the Large Hadron Collider?

The LHC will be upgraded between 2025 and 2027. Scientists will collaborate to plan new investigations in the tunnel. They hope to make protons collide more frequently. Thousands of scientists worldwide will analyse the extra data. They expect many new discoveries.

▲ Archana Sharma of India worked on the Higgs boson team.

◀ Analysing data at the Large Hadron Collider.

Questions

1. Explain why peer review is important.
2. Give two advantages of collaboration in science.
3. Suggest one advantage and one disadvantage of working in an international team.
4. Suggest why it is easier for scientists to collaborate in international teams now than it was 100 years ago.

Key points

- Scientists build on others' work to develop scientific knowledge over time.
- Scientists, who are expert in different things, collaborate.
- In peer review, experts check scientific papers to see if they are good enough to publish.

Extension 7.5

Proton and nucleon numbers

Objectives

- Define the terms *proton number*, *nucleon number* and *mass number*
- Deduce the proton and nucleon numbers of atoms of different elements
- Explain what isotopes are

Proton number

Every atom of hydrogen has one proton in its nucleus. Every atom of oxygen has eight protons in its nucleus. Every atom of magnesium has 12 protons in its nucleus.

The number of protons in the nucleus of an atom is its **proton number**. The table gives the proton numbers of some elements. Every atom of a certain element has the same proton number.

In modern periodic tables, elements are arranged in order of proton number.

Element	Proton number
hydrogen	1
oxygen	8
magnesium	12

Atoms are neutral. This is because an atom has an equal number of protons and electrons. So the proton number of an element tells you the number of protons, which is also the number of electrons in one atom of the element.

Key
Proton number
atomic symbol
name
Nucleon number

1	2	3	4	5	6	7	8	9	10	11	12	13	14	15	16	17	18
1 H hydrogen 1																	2 He helium 4
3 Li lithium 7	4 Be beryllium 9											5 B boron 11	6 C carbon 12	7 N nitrogen 14	8 O oxygen 16	9 F fluorine 19	10 Ne neon 20
11 Na sodium 23	12 Mg magnesium 24											13 Al aluminium 27	14 Si silicon 28	15 P phosphorus 31	16 S sulfur 32	17 Cl chlorine 35.5	18 Ar argon 40
19 K potassium 39	20 Ca calcium 40	21 Sc scandium 45	22 Ti titanium 48	23 V vanadium 51	24 Cr chromium 52	25 Mn manganese 55	26 Fe iron 56	27 Co cobalt 59	28 Ni nickel 59	29 Cu copper 64	30 Zn zinc 65	31 Ga gallium 70	32 Ge germanium 73	33 As arsenic 75	34 Se selenium 79	35 Br bromine 80	36 Kr krypton 84
37 Rb rubidium 85	38 Sr strontium 88	39 Y yttrium 89	40 Zr zirconium 91	41 Nb niobium 93	42 Mo molybdenum 96	43 Tc technetium –	44 Ru ruthenium 101	45 Rh rhodium 103	46 Pd palladium 106	47 Ag silver 108	48 Cd cadmium 112	49 In indium 115	50 Sn tin 119	51 Sb antimony 122	52 Te tellurium 128	53 I iodine 127	54 Xe xenon 131
55 Cs caesium 133	56 Ba barium 137	57–71 lanthanoids	72 Hf hafnium 178	73 Ta tantalum 181	74 W tungsten 184	75 Re rhenium 186	76 Os osmium 190	77 Ir iridium 192	78 Pt platinum 195	79 Au gold 197	80 Hg mercury 201	81 Tl thallium 204	82 Pb lead 207	83 Bi bismuth 209	84 Po polonium –	85 At astatine –	86 Rn radon –
87 Fr francium –	88 Ra radium –	89–103 actinoids	104 Rf rutherfordium –	105 Db dubnium –	106 Sg seaborgium –	107 Bh bohrium –	108 Hs hassium –	109 Mt meitnerium –	110 Ds darmstadtium –	111 Rg roentgenium –	112 Cn copernicium –	113 Nh nihonium –	114 Fl flerovium –	115 Mc moscovium –	116 Lv livermorium –	117 Ts tennessine –	118 Og oganesson –

Note: This periodic table does not include all the elements.

▲ *A larger periodic table can be found at the back of this book.*

Nucleon number

Protons and neutrons make up the nucleus of an atom. Particles in the nucleus – protons and neutrons – are called **nucleons**. The total number of protons and neutrons in an atom is its **nucleon number**. For example, an atom of oxygen has 8 protons and 8 neutrons. Its nucleon number is 16.

The nucleon number of an atom is also called its **mass number**. It gives the relative mass of an atom compared to other atoms.

The table gives the numbers of protons and neutrons in some atoms. It also shows their nucleon numbers.

Atom of the element	Number of protons	Number of neutrons	Nucleon number
fluorine	9	10	(9 + 10) = 19
magnesium	12	12	(12 + 12) = 24
argon	18	22	(18 + 22) = 40

If you know the proton number and the nucleon number of an atom, you can work out how many protons and neutrons are in its nucleus.

Worked example

An atom of boron has a proton number of 5 and a nucleon number of 11. How many protons and neutrons does it contain?

The proton number shows that the number of protons in a boron atom is 5.

The number of neutrons = nucleon number – proton number
= 11 – 5
= 6

Isotopes

Every carbon atom has 6 protons. Every carbon atom also has 6 electrons. Most carbon atoms have 6 neutrons. However, some carbon atoms have 8 neutrons, and some have 7 neutrons.

Atoms of the same element that have different numbers of neutrons are called **isotopes**. The different isotopes of an element have different nucleon numbers. The table shows the nucleon numbers of the three isotopes of carbon.

Atom of the element...	Number of protons	Number of neutrons	Nucleon number
carbon	6	6	12
carbon	6	7	13
carbon	6	8	14

Every chlorine atom has 17 protons and 17 electrons. About 75% of chlorine atoms have 18 neutrons. The other 25% of chlorine atoms have 20 neutrons. There are two isotopes of chlorine. The nucleon number of one isotope is (17 + 18) = 35. The nucleon number of the other isotope is (17 + 20) = 37.

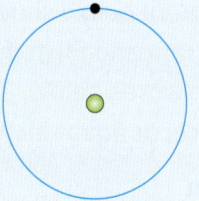
hydrogen-1
(1 proton)

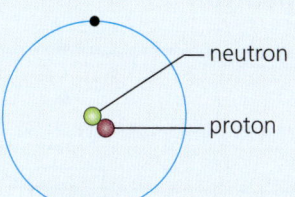
hydrogen-2
(1 proton and 1 neutron)

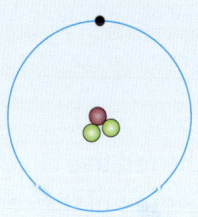
hydrogen-3
(1 proton and 2 neutrons)

▲ *Hydrogen has three isotopes.*

Key points

- Proton number is the number of protons in an atom of an element.
- Nucleon number is the total number of protons and neutrons in an atom of an element.
- Isotopes are atoms of the same element with different numbers of neutrons.

Questions

1. Write definitions for *proton number* and *nucleon number*.
2. An atom of an element has 15 protons and 16 neutrons. Give its proton number and its nucleon number.
3. An atom has a proton number of 19 and a nucleon number of 39. Give the number of protons and neutrons in the atom.

Review 7.6

1. The diagram below shows a neutral atom. Copy the diagram and fill in the missing labels. Choose from the words below. Use each word once, more than once, or not at all. [3]

   electron
   proton
   nucleus
   neutron

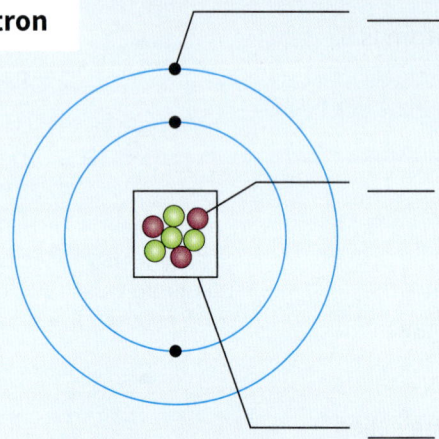

2. What is the relative charge on a proton? Write the letter of the correct answer. [1]

   A −1
   B 0
   C +1
   D $\frac{1}{2000}$

3. What is the relative charge of a neutron? Write the letter of the correct answer. [1]

   A −1
   B 0
   C +1
   D $\frac{1}{2000}$

4. What is the relative mass of a neutron? Write the letter of the correct answer. [1]

   A 2
   B 1
   C 0
   D $\frac{1}{2000}$

5. What type of force holds the nucleus and electrons together? Write the letter of the correct answer. [1]

   A electrostatic
   B friction
   C gravity
   D magnetic

6. Copy and complete the table to show the relative mass and charge of each sub-atomic particle. [4]

Sub-atomic particle	Charge	Relative mass
proton	+1	
neutron		
electron		$\frac{1}{2000}$

7. The picture shows some sulfur.

   Each sulfur atom has 16 protons.

   a. Name the part of the atom that the protons are in. [1]

   b. Give the number of electrons in a sulfur atom. [1]

   c. Name the type of force that holds the protons and electrons together. [1]

8. The picture shows some mercury.

a. Is mercury a metal or non-metal? Use the periodic table to help you decide. [1]

b. A mercury atom has 80 protons and 120 neutrons.

   i. Calculate the total number of sub-atomic particles in the nucleus of a mercury atom. [2]

   ii. Deduce the number of electrons in a mercury atom. [1]

   iii. Explain why a mercury atom is electrically neutral. [1]

9. The table shows the numbers of protons and neutrons in the atoms of some elements.

Atom of…	Number of protons	Number of neutrons
boron	5	6
helium	2	2
lithium	3	4

Some students made some models of the atoms. They used eggs, tomatoes, grapes, nuts, and dried fruit. The models show protons, neutrons, and electrons.

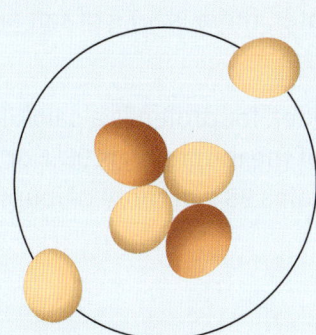

Model A

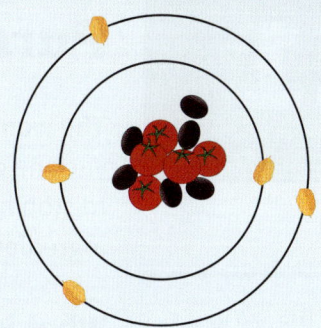

Model B

Model C

a. Give the number of electrons in a boron atom. Choose data from the table to help you to answer this question. [1]

b. Give the number of electrons in model C. [1]

c. Give the colour of the protons in model B. Use the table and the picture to help you answer this question part. [1]

d. Give the letter of the model of a boron atom. [1]

e. Give the letter of the model of a helium atom. [1]

f. Describe **two** ways in which model C represents the sub-atomic particles better than model A. [2]

g. Another student wants to use eggs to make a model of a beryllium atom. The atom has 4 protons and 5 neutrons. Give the total number of eggs she would need. [2]

8.1 Pure substances

Objectives

- Define the terms *pure substance* and *purity*
- Describe how to show that a substance is pure
- Make a conclusion from a graph
- Describe one application of science

The woman is being vaccinated. The vaccine contains an active ingredient to stop her getting ill. But the main ingredient is water.

What is a pure substance?

In science, a **pure substance** consists of one substance only. There is nothing mixed with it. A pure substance can be an element or a compound.

The vaccine company uses pure water in its vaccines. Pure water contains water molecules only.

The vaccine company does not use tap water. Tap water is not pure. It has other substances mixed with it, such as chlorine and carbonates.

▲ Water is a compound. Pure water contains water molecules only. Not to scale.

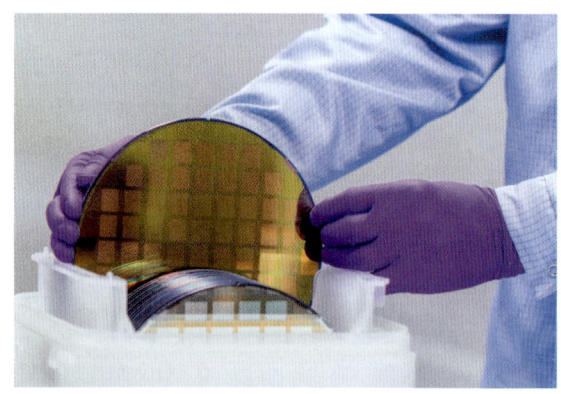

▲ Silicon is an element. Pure silicon contains silicon atoms only. It is used to make integrated circuits in phones.

▲ The word *pure* has different meanings in science and in everyday life. In everyday life, a pure substance is one that has not been processed.

▲ The Arabian Sea

What is purity?

Purity describes how much of a substance is in a mixture.

The Arabian Sea is salty. In 100 g of seawater, there is 3.7 g of salt and 96.3 g of water. The purity of the water is 96.3%.

A chemical company sells ibuprofen powder. Ibuprofen is a painkiller. Its chemical formula is $C_{13}H_{18}O_2$. The purity of the ibuprofen is 98% or better. This means that 100 g of the powder contains 98 g or more of ibuprofen.

◀ Ibuprofen powder

Pure substances and solutions

Thinking and working scientifically

Is it pure?

The temperature of a pure substance does not change while it melts or freezes.

Farooq investigates how stearic acid cools. The formula of stearic acid is $C_{18}H_{36}O_2$. He pours hot liquid stearic acid into a test tube and lets it cool. He records the temperature every minute. His results are in the table.

Farooq plots his data on a graph. He makes a conclusion:

The temperature of the stearic acid does not change while it freezes. This shows that the stearic acid is pure.

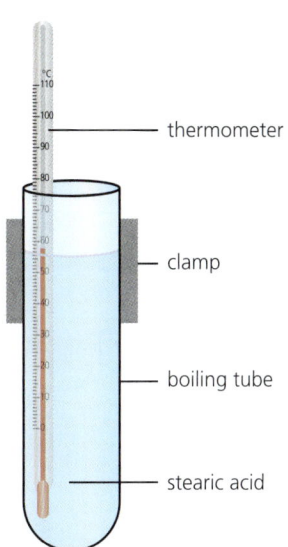

▲ Apparatus to investigate freezing

Time (min)	0	1	2	3	4	5	6	7	8	9	10
Temperature (°C)	96	77	70	70	70	70	70	66	63	61	58

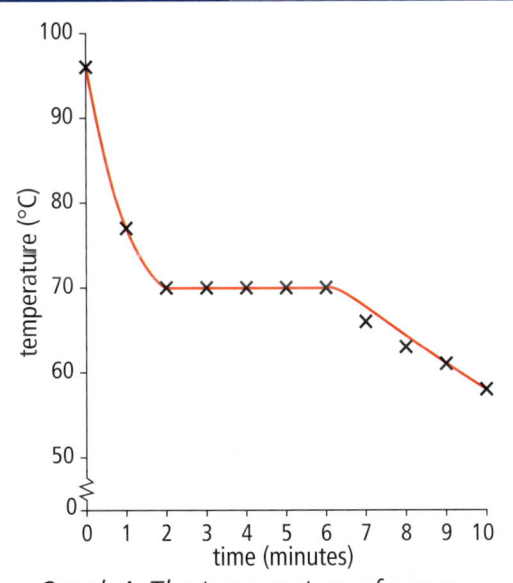

▲ Graph A: The temperature of a pure substance does not change as it freezes.

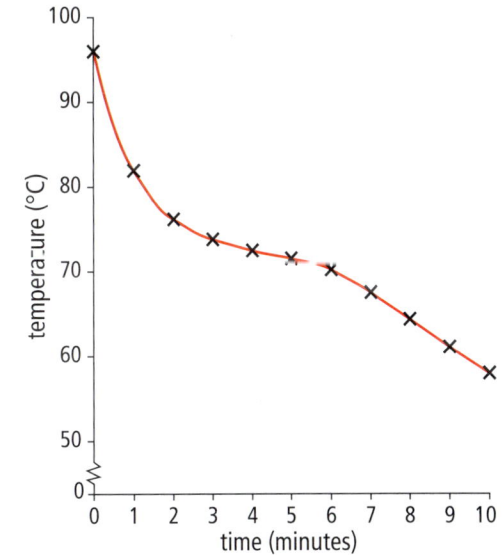

▲ Graph B. The temperature of an impure substance changes as it freezes.

Questions

1. Write definitions for *pure substance* and *purity*.
2. Explain why seawater is not a pure substance.
3. **TWS** Use data from Graph A to give the melting point of stearic acid.
4. A company supplies paracetamol powder with a purity of 99%. Calculate the mass of paracetamol in 1000 g of the powder.
5. **TWS** Find the formulae of ibuprofen and stearic acid on this page. Name the three elements whose atoms are in the compounds.

Key points

- A pure substance consists of one element or compound only.
- Purity describes how much of a substance is in a mixture.

127

Science in context 8.2

Objectives

- Define the term *desalination*
- Describe some benefits and problems of desalination

Drinking seawater

It rains very little in Saudi Arabia. There are no permanent rivers, and only one lake. So where does the nation get its water?

▲ *Riyadh, Saudi Arabia*

▲ *Jebel Fihrayn, Saudi Arabia*

Obtaining pure water from seawater

About half of the drinking water in Saudi Arabia comes from the sea. A process called **desalination** removes salt from the water. The process also removes other dissolved substances. This makes pure water, which is safe to drink.

Obtaining pure water in the lab

There are different methods of removing salt from seawater. In the lab, you can use distillation.

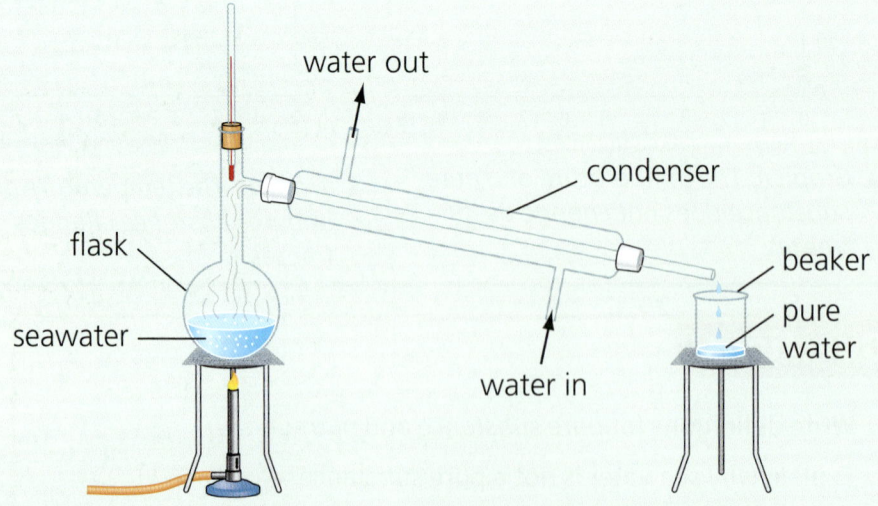
▲ *Distillation apparatus*

The seawater boils. This makes steam. Steam travels through the condenser. It cools and condenses. Pure, liquid water drips into the beaker. The salt remains in the flask.

Pure substances and solutions

Pure water for millions

Scientists and engineers apply their science knowledge – and creativity – to devise processes to separate pure water from seawater on a big scale. The processes are used in Saudi Arabia and other countries.

- One process is similar to laboratory distillation. The pressure is reduced so that the water boils at a lower temperature. This makes it cheaper.
- Another process is called reverse osmosis. This works a bit like filtering.

▲ This desalination factory is in Dubai, United Arab Emirates. It makes pure water from seawater.

Desalination problems

Desalination makes drinking water for millions. But it has disadvantages:

- The water is completely pure. Some people do not like its taste.
- The pure water does not contain dissolved calcium compounds. Other sources of drinking water may contain these compounds, which benefit health.
- Some methods of desalination need huge amounts of electricity.
- Sea animals and plants may die where very salty water returns to the sea.

▲ Ships and submarines produce pure water by desalination, too.

Questions

1. Explain why desalination is an important source of drinking water in some countries.
2. Water obtained by desalination is pure. Write the definition for *pure* in science.
3. Suggest some benefits and problems of obtaining drinking water from seawater.

Key points

- Desalination removes salt from seawater.
- Desalination is an important source of drinking water, but there are problems linked to the process.

8.3 Chromatography

Yadira has some sweets. She wants to know what colourings are in them. How can she find out?

Objectives

- Describe how chromatography separates and identifies substances in a mixture
- Describe a useful scientific enquiry

What is chromatography?

Chromatography is a method used to separate and identify the substances in a mixture. It also shows if a substance is pure. Chromatography works if all the substances in a mixture dissolve in one solvent.

Chromatography of ink

Radi sets up the apparatus on the left. The water moves up the paper. It takes the dyes in the ink with it. Different dyes move at different speeds, so they separate. This makes a chromatogram.

The chromatogram below the apparatus shows that green felt tip pen ink is a mixture of two dyes – yellow and blue. The blue dye travels further.

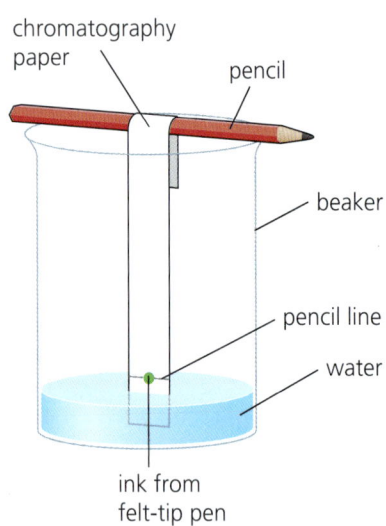

Chromatography of sweets

Yadira makes a chromatogram with five sweets. The chromatogram shows that the brown dye is a mixture of red, yellow, and blue dyes. The red dye makes one spot. It is a pure substance.

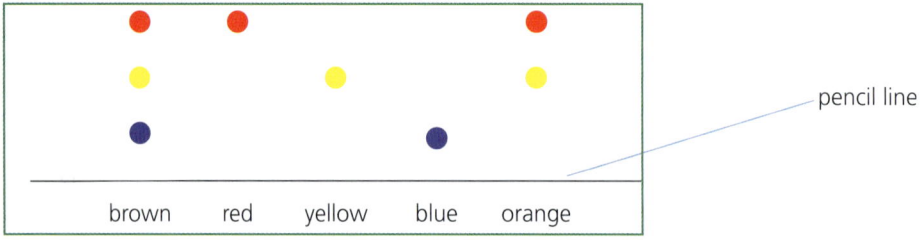

▲ This chromatogram shows that the brown dye is a mixture.

▲ A chromatogram from a green pen

> ### 🔬 Science in context
> #### Using chromatography
> Paper chromatography is useful for separating substances in ink, plants, and sweets. There are other types of chromatography. They all separate mixtures. They are useful to identify substances, to measure amounts, and to check purity. For example:
> - Scientists and pharmaceutical companies use chromatography in the production of some COVID-19 vaccines.
> - Airport security officers use chromatography to test clothes and luggage for explosives.
> - Scientists use chromatography to identify food nutrients.

Pure substances and solutions

Thinking and working scientifically

Chromatography enquiry

In Africa and Asia, up to half a million children go blind each year. One cause is a lack of vitamin A.

Many people eat large amounts of cassava. Nigerian scientist Steve Adewusi and Australian Howard Bradbury wondered if different types of cassava have different amounts of vitamin A. Would switching to vitamin A rich cassava prevent blindness?

The scientists asked a scientific question:

How much vitamin A is in the roots and leaves of different types of cassava?

▲ *Yellow cassava root*

▲ *White cassava root*

Carrying out the investigation

The scientists collected evidence. They used chromatography to measure the amounts of vitamin A in yellow and white cassava roots. They found the amounts of vitamin A in dark green and light green leaves.

The scientists also collected secondary data – the amount of vitamin A needed to be healthy – from the World Health Organization (WHO). The scientists trusted the data because WHO is well respected.

▲ *Cassava leaf stew*

Analysing the evidence

The scientists studied their results. Here is part of their conclusion:

Dark green leaves have more vitamin A than light green leaves.

The scientists calculated the amount of cassava that has enough vitamin A to be healthy to eat. They worked out that eating yellow cassava root, and cassava leaves, can provide enough vitamin A. White roots alone cannot.

The scientists made a recommendation. To prevent blindness, children should eat yellow cassava, not white. They should also eat cassava leaves.

Questions

1. Write down three purposes of doing chromatography experiments.
2. Look at the chromatogram of the sweets.
   a. Identify the two dyes that are mixed in orange sweets.
   b. Give the number of dyes in the red sweets.
   c. Give the colours of three dyes that are pure substances.

Key points

- Chromatography separates and identifies substances that dissolve in the same solvent.

8.4 Solutions and concentration

Objectives

- Define the terms *concentration*, *dilute*, and *concentrated*
- Use the particle model to explain concentration
- Describe how emergency workers apply science

The lorry is carrying ethanoic acid. The hazard symbol shows that the acid is corrosive. It could burn your eyes and skin. But ethanoic acid in vinegar is safe to eat in pickles. What's the difference?

Concentrated or dilute?

In vinegar, ethanoic acid is mixed with a large amount of water. This is a **dilute** solution. But the ethanoic acid in the lorry is mixed with little water. It is a **concentrated** solution. Concentrated acids are more corrosive than dilute ones.

What is concentration?

As you know, a solution is a mixture that forms when a solute dissolves in a liquid. The liquid is the solvent, and the substance that dissolves is the solute.

The **concentration** of a solution is a measure of the number of solute particles in a volume of solution.

▲ There is the same volume of solution in both cups. There are more tea particles in cup B. The solution in cup B has the higher concentration.

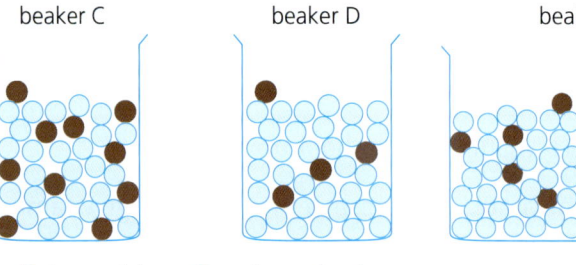

▲ There is the same volume of solution in both beakers. There are more tea particles in beaker C. The solution in beaker C has the higher concentration. Not to scale.

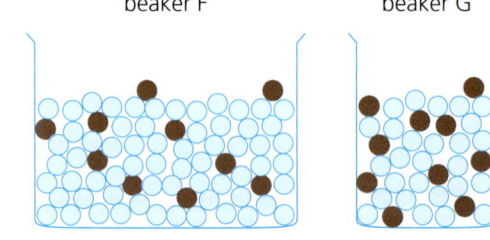

▲ Both beakers have the same number of tea particles. Beaker G has more tea particles in a certain volume. The solution in beaker G has the higher concentration. Not to scale.

🧪 Thinking and working scientifically

Concentration models: strengths and limitations

The diagrams above model concentrated and dilute solutions. The strength of these models is that they show that a concentrated solution has more solute particles in a certain volume. Their main limitation is that they are not to scale. There are millions and millions and millions of water and solute particles in a cup or beaker.

Pure substances and solutions

 Science in context

Acid spill

A lorry is carrying many litres of concentrated sulfuric acid. The acid is corrosive. Its formula is H_2SO_4. Suddenly, acid starts leaking onto the road. The lorry stops. Fire fighters arrive. They are wearing special clothes, to protect their skin and eyes.

The firefighters add water to the acid on the road. The water dilutes the acid. The firefighters are applying their science knowledge to make the area safe.

▲ The back of a lorry carrying concentrated sulfuric acid.

▲ Adding water to dilute acid on the road.

Questions

1. Write the definition for *concentration*.
2. Copy the particle diagram of a solution. Then draw a diagram of a more concentrated solution.

3. Suggest why firefighters add water to spilled concentrated acid, not an alkali.
4. Find the formula of sulfuric acid on this page. Name the three elements whose atoms are in the compound.

Key points

- Concentration is a measure of the number of solute particles in a volume of solution.
- A dilute solution is mixed with a large amount of water. There are few solute particles in a certain volume.
- A concentrated solution is mixed with little water. There are many solute particles in a volume.

Thinking and working scientifically

8.5

Objectives

- Give the definitions of *risk* and *hazard*
- Describe how to plan, carry out and analyse an investigation

How much salt is in the sea?

Emebet lives near the Red Sea. She knows that seawater is a solution of salt, sodium chloride (NaCl). Emebet wonders how much salt is in the Red Sea. She does an investigation.

▲ The arrow points to the Red Sea.

▲ A beach on the Red Sea

Planning the investigation

Emebet turns her idea into a scientific question:

What mass of salt is dissolved in $20\,cm^3$ of Red Sea water?

She makes a plan:

- Measure out $20\,cm^3$ of seawater.
- Heat until all the water evaporates.
- Measure the mass of salt that remains.

As you know, there are different types of investigation. Emebet cannot answer her question by doing a fair test, because she is not investigating the effect of one variable on another.

Risk assessment

Emebet thinks about the hazards. A **hazard** is a possible source of danger or damage. The hazards are hot apparatus and salt spitting from boiling seawater.

Emebet must reduce the **risks** from these hazards. She writes a risk assessment.

Hazard	Risk from hazard	Reduce chance of injury and damage by...
hot apparatus	burns to skin	- turning off the Bunsen burner as soon as the water has evaporated - waiting for the equipment to cool before touching it - standing back while heating
salt spitting from boiling seawater	burns to skin and damage to eyes	wear eye protection

Pure substances and solutions

Carrying out the investigation

Choosing and using equipment

Emebet pours seawater into a measuring cylinder. She says its volume is 20 cm³.

A friend tells Emebet to add more seawater. The bottom, not the top, of its curved surface should be level with the 20 cm³ mark.

Next, Emebet finds the mass of the empty evaporating basin. It is 25 g. Emebet notices a mistake. The evaporating basin is wet, so its measured mass is too high. Emebet dries the evaporating basin. She places it on the balance again. Its mass is 25 g.

Emebet is surprised that the mass values are the same for the dry and wet basins. She needs a balance that measures smaller differences in mass. She uses a different balance to find the mass of the dry evaporating basin.

Mass of evaporating basin = 25.10 g

Presenting evidence

Emebet organises the data in a table. This will make it easier to use the data in calculations. There is a unit in the column heading.

	mass (g)
mass of empty evaporating basin	25.10
mass of evaporating basin + salt	25.80
mass of salt	

Emebet heats the seawater. The water evaporates. Salt remains. She measures the mass of the evaporating basin and salt.

Analysing the evidence

Analysis and conclusion

Emebet calculates the mass of salt:

mass of salt = (mass of evaporating basin + salt) − (mass of evaporating basin)

= 25.80 g − 25.10 g

= 0.70 g

The calculation is clearly set out. This makes it easy to check, or do a similar calculation later.

To finish, Emebet uses a secondary source to find the volume of seawater in the Red Sea. She uses this value, and her investigation data, to estimate the mass of salt in the Red Sea. She writes a conclusion.

Evaluation

Emebet suggests two improvements:
- Repeat the investigation three times to obtain **reliable** data (data that she is confident are correct).
- Calculate the mean mass of salt in the three repeats. The mean is probably more **accurate** (closer to the actual value) than a single result.

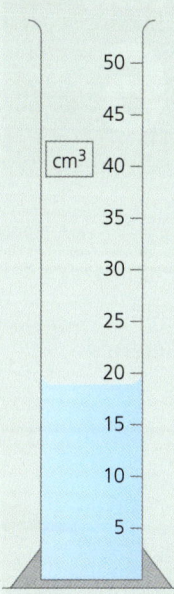

▲ Emebet uses a measuring symbol to measure the volume.

Key points

- An investigation involves planning, carrying out, and analysis.

Questions

1. Write definitions for *hazard* and *risk*.
2. Estimate the volume of water in the measuring cylinder in the diagram.
3. Explain why Emebet suggests doing the investigation three times.
4. The volume of water in the Red Sea is approximately 200 000 km³. The mass of salt in 1 km³ is 35 000 000 000 kg. Estimate the mass of salt in the Red Sea.

135

Science in context 8.6

Chlorine and water

Dirty water kills. The World Health Organization estimates that diseases spread by untreated water kill nearly half a million people every year. The diseases include cholera, typhoid, and dysentery.

Objective

- Evaluate an issue that requires science understanding

Life-saving chlorine

Adding chlorine to water destroys the bacteria, viruses, and parasites that cause these diseases. Compounds of chlorine, such as calcium chloride or sodium hypochlorite, do the same. These substances save many lives.

A dangerous by-product?

River and lake water contain dissolved substances from the soil. When chlorine is added, it reacts with these substances. The chemical reactions make a mixture of products, called trihalomethanes (THMs). THMs include compounds with these formulae: $CHCl_3$, $CHBrCl_2$, $CHClBr_2$, and $CHBr_3$.

Many scientists have investigated whether, over many years, THMs in tap water increase the risk of cancer. Different groups of scientists made different conclusions. Almost all say that, if there is an increased risk of cancer, it is tiny.

Evaluating the issue

Water company workers know that chlorinating water saves lives. They also know about the risk of adding THMs. They use science to help them decide what to do. Almost all water companies chlorinate tap water. They judge that the benefits outweigh the risks.

▲ Adding chlorine or its compounds to water destroys disease-causing microbes.

🧪 Thinking and working scientifically

Investigating the effects of chlorinating water

As you know, many groups of scientists have asked the question:

What are the health effects of drinking chlorinated water?

Here is how Rabi Sharma and Sudha Goel, from West Bengal, India, investigated the question.

Carrying out the investigation

The scientists collected evidence from three groups:

- Group 1 had been drinking chlorinated water for 30 years or more.
- Group 2 had little access to chlorinated water at home.
- Group 3 had no access to chlorinated water at home.

The scientists asked many people in one city to join the study, and 1810 people joined in. The big sample size helps to make the evidence reliable.

The scientists did not collect evidence from people younger than 30. They wanted to investigate people who had been drinking water for 30 years.

Interview evidence

The scientists collected evidence by visiting people in their homes. They asked questions about:

- how long they had lived in their home
- their age
- their education
- their income (how much money they earn)
- their health
- if they filter or boil tap water before drinking.

▲ *Collecting water.*

The scientists asked the first two questions to check whether to include someone in the investigation. They asked about education and income because these two variables affect health.

The scientists asked questions about health. They asked about cancer. They asked about waterborne diseases, such as cholera, typhoid, and dysentery. They asked about skin infections and kidney disease.

Experimental evidence

The scientists measured the concentrations of THMs at five places in the city water supply. They also found that almost everyone in the investigation filtered or boiled their water before drinking it.

Analysing the evidence

Making a conclusion

The scientists thought about their evidence. They used their data to do calculations. They made two conclusions:

- Drinking chlorinated water does not significantly increase the risk of cancer.
- Drinking chlorinated water reduces the risk of cholera, typhoid, and dysentery. It also reduces the risk of skin infections and kidney disease.

Reporting the investigation

The scientists wrote a report and sent it to a scientific journal. Other scientists checked the report. It was published.

The large number of people in the study, and the care with which it was done, mean that other scientists will trust the conclusions.

Questions

1. Describe the benefits of chlorinating drinking water.
2. Describe one possible risk of chlorinating drinking water.
3. The scientists collected data from 1810 people. Suggest why they did not collect data from an even bigger number of people.

Key points

- Scientific knowledge helps individual people, organisations, and governments to make health decisions.

8.7 Solubility

How much sugar is dissolved in a 330 cm³ bottle of cola? The answer is about 40 g. That's 10 teaspoonfuls.

Objectives

- Define *solubility* and *saturated solution*
- Describe how solubility varies with temperature
- Identify trustworthy sources of secondary data
- Make conclusions from data

What is a saturated solution?

At room temperature, you can dissolve more than 200 g of sugar in 100 g of water. If you continue to add more sugar, it does not dissolve. It falls to the bottom. This is a saturated solution.

A **saturated solution** contains the maximum mass of solute that will dissolve. There is always some undissolved solute in the container.

▲ A bottle of cola

What is solubility?

The maximum mass of a substance that dissolves in 100 g of water is its **solubility**. Every substance has its own solubility. For example:

Substance	Solubility at 20 °C (g/100 g of water)
sugar	202
salt	36

The data show that more sugar than salt can dissolve in 100 g of water. Sugar is more soluble than salt.

▲ Different substances have different solubilities.

Thinking and working scientifically

Data from secondary sources

Laziz wants to compare the solubility of different substances. He cannot do an experiment because he does not have a balance to measure mass.

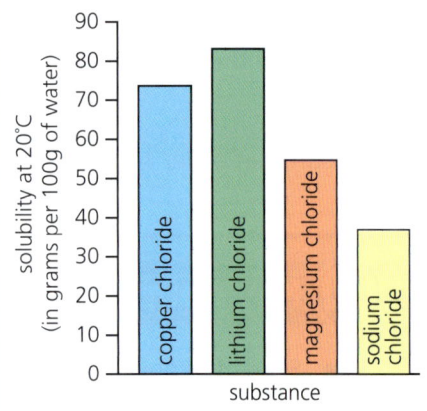

Laziz collects data from a secondary source. As you know, secondary sources provide evidence you have not collected yourself in an investigation. They include books, scientific journals, and the Internet. If another student gives you evidence from their investigation, the student is a secondary source.

You can trust some secondary sources more than others. Scientific journals are trustworthy, because they are written and checked by experts. You can also trust data books. The quality of evidence from the Internet varies. Evidence on a chemistry website, such as the International Union of Pure and Applied Chemistry (IUPAC), is more trustworthy than evidence on social media.

Pure substances and solutions

> **Making conclusions from data**
>
> Laziz finds the data he needs. He plots the data on a bar chart. He chooses a bar chart because the independent variable – the substance – is categoric.
>
> You can practise making conclusions by answering questions 2, 3, and 4.

How does solubility change with temperature?

Can you dissolve more sugar in hot water, or in cold water?

The data show that the higher the temperature, the higher the solubility. More sugar dissolves in hot water than in cold water.

Most substances get more soluble as temperature increases. But the difference is greater for some substances than others. Compare the solubility values of sugar and salt at 20 °C and 100 °C.

Temperature (°C)	Solubility of sugar (g/100 g of water)
20	202
40	236
60	289
80	365
100	476

▲ Table 1

Temperature (°C)	Solubility of salt (g/100 g of water)
20	36
100	39

▲ Table 2

The graph shows how the solubilities of six more substances change with temperature. Line graphs are chosen because both variables are continuous.

The graph shows that, for all substances except cerium(III) sulfate, solubility increases with temperature.

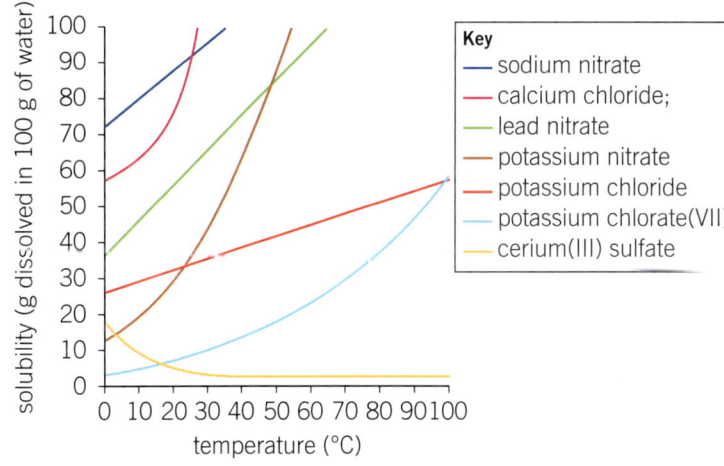

Questions

1. Write the definitions for *solubility* and *saturated solution*.
2. Use the bar chart opposite to identify:
   a. the most soluble substance shown
   b. the least soluble substance shown
   c. the solubility of copper chloride.
3. Use data in Table 1 to describe how the solubility of sugar changes with temperature.

4. Use the line graphs to compare how the solubilities of potassium nitrate and cerium(III) sulfate change with temperature.

Key points

- Solubility is the maximum mass of a substance that dissolves in 100 g of water.
- For most substances, solubility increases with temperature.
- Trustworthy sources of secondary data include scientific journals, data books, and some websites.

139

Thinking and working scientifically

8.8

Investigating solubility and temperature – doing an investigation

Objective

- Describe how to plan and carry out an investigation

Do you add sugar to hot drinks? How can you investigate dissolving at different temperatures?

Planning an investigation

Making a prediction

Hani is investigating the scientific question:

How does water temperature affect the mass of sodium carbonate that dissolves?

Hani knows that, for most substances, solubility increases with temperature. He thinks that sodium carbonate might behave in the same way. He makes a prediction:

As the temperature of water increases, the mass of sodium carbonate that dissolves will increase.

Considering variables

A variable is anything that might affect what happens in an investigation. Hani lists the possible variables in his investigation:

- water temperature
- water volume
- mass of sodium carbonate that dissolves.

Hani makes decisions about the variables:

- independent variable (variable to change) – water temperature
- dependent variable (variable to measure) – mass of sodium carbonate that dissolves.

Hani wants to do a fair test, so he keeps the water volume the same. This is the control variable.

Choosing equipment

Hani plans what to do. He will measure out 100 cm³ of water and cool it to 0 °C. He will add sodium carbonate, 1 g at a time, and stir until no more dissolves. He will then repeat at different temperatures.

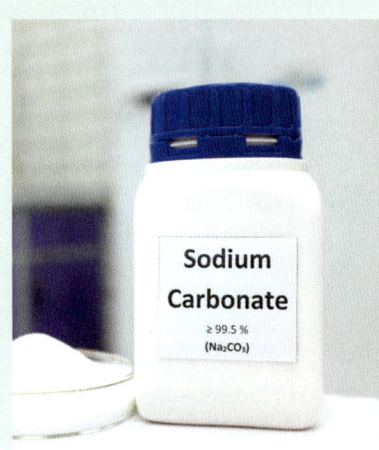

▲ Sodium carbonate in a laboratory

Pure substances and solutions

Hani needs to choose suitable apparatus.

Hani could use a measuring cylinder or a beaker to measure the volume of water. He chooses a measuring cylinder because it measures smaller differences in volume.

Hani has a choice of two thermometers. The clinical thermometer measures temperatures between 35 °C and 42 °C. Hani wants to measure over a wider range of temperatures, so he chooses a thermometer with a measurement range of 0 °C to 100 °C.

Hani's school has two instruments to measure mass. He can use the balance with the balance weights to measure mass changes as small as 1 g. The electric balance measures smaller differences in mass. Hani decides to use the balance with weights because the electricity supply is not reliable.

Hani uses a Bunsen burner to heat the water. He uses a stirring rod to stir the mixtures.

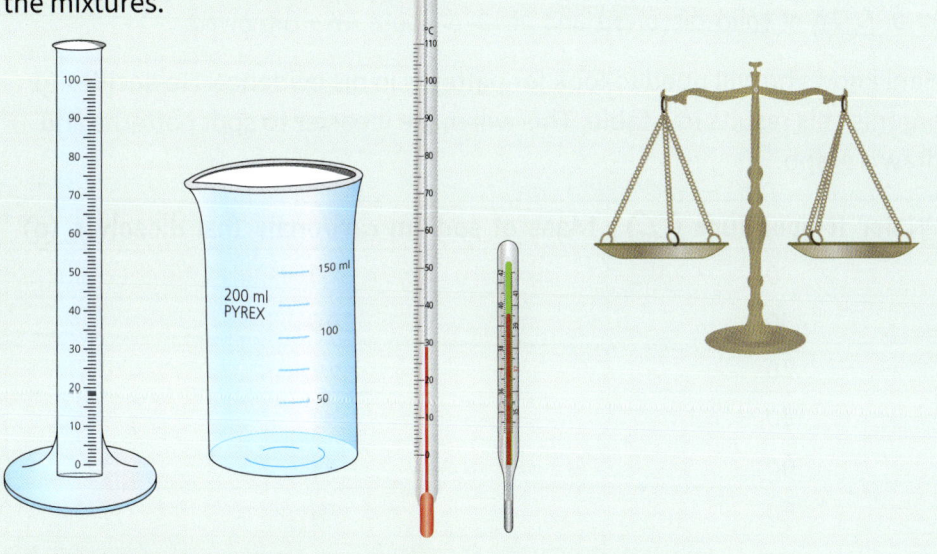
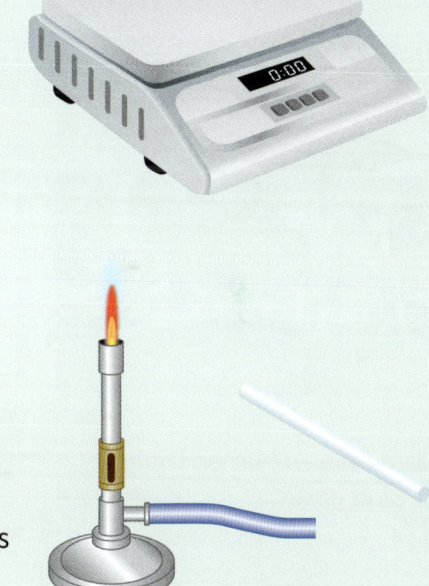

▲ The apparatus that Hani can choose from. Not to scale.

Carrying out the investigation

Hani uses his apparatus to find the mass of sodium carbonate that dissolves in water at six temperatures. His results are on the next page.

Questions

1. Define the terms *independent variable* and *dependent variable*.
2. Write down the chemical formula of sodium carbonate. Use the photo of sodium carbonate to help you.
3. Explain why Hani keeps one variable constant.
4. Explain why Hani chose a measuring cylinder, not a beaker, to measure water volume.
5. Write the column headings for a table that Hani could write his results in.

Key points

Planning an investigation involves:
- making a prediction
- considering variables
- identifying evidence to collect
- choosing equipment.

141

Thinking and working scientifically

8.9

Investigating temperature and solubility – writing up an investigation

Objective

- Describe how to present evidence, write conclusions, and evaluate investigations

Hani is investigating the scientific question:

How does water temperature affect the mass of sodium carbonate that dissolves?

Carrying out the investigation

Recording data

Hani begins to collect evidence. He writes his observations on a scrap of paper:

At 0 °C, 7g of solid dissolved and at 10 °C about 48g dissolved.

Hani knows he will need to look for patterns in his evidence. He decides to organise his results in a table. This will make it easier to spot patterns and draw a graph.

Water temperature (°C)	Mass of sodium carbonate that dissolves (g)
0	7
10	48
20	22
30	39
40	49
50	57

▲ A Bunsen burner, tripod, and gauze

When you draw a results table:

- write the independent variable (variable to change) in the left column
- write the dependent variable (variable to observe or measure) in the right column
- include units in the column headings.

Analysing the evidence

Bar chart or line graph?

Hani studies the data in the table. He notices that, as temperature increases, so does the mass of sodium carbonate that dissolves. His prediction is correct.

He highlights an anomalous result, which does not fit the pattern. He realises that the water was hotter than 10 °C when he took this reading.

Hani decides to examine the pattern more closely. He wants to draw a bar chart or line graph. Which is more suitable?

- Draw a bar chart when the independent variable is categoric or discrete:
  - categoric variables are described by words
  - discrete variables are numbers with no in-between values.
- Draw a line graph when the independent and dependent variables are continuous. A continuous variable can have any numerical value.

Drawing a line graph

Hani plots his data on a graph. When you draw a line graph:

- Label the *x*-axis with the name and units of the independent variable.
- Label the *y*-axis with the name and units of the dependent variable.
- Choose a scale for each axis.
- Write values on the lines on the *x*-axis – use evenly spaced numbers.
- Write values on the lines on the *y*-axis – use evenly spaced numbers.

Then draw a line of best fit to show the pattern. The line of best fit can be a straight line or a curve. Hani decides that the line of best fit is a straight line. He ignores the anomalous result.

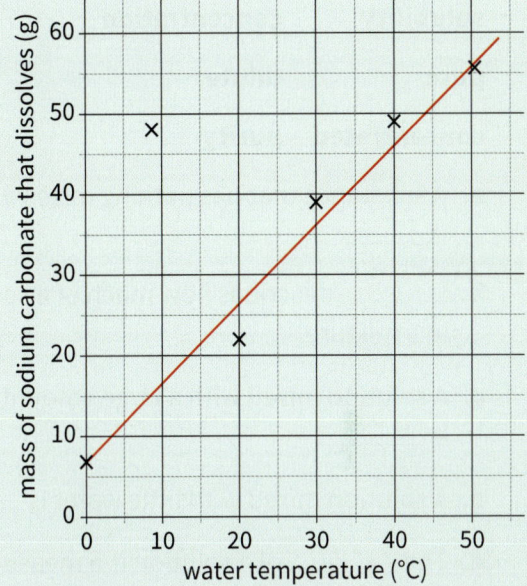

Making a conclusion

Hani makes a conclusion:

The solubility of sodium carbonate increases with temperature. My prediction was correct.

His teacher suggests adding more detail to the conclusion:

The solubility of sodium carbonate increases steadily from 7 g/100 g of water at 0 °C to about 50 g/100 g of water at 50 °C. My prediction was correct.

If Hani had learnt about a reason for the pattern, he should include it.

Evaluating the investigation

Hani says that a limitation of his investigation is that he did not find the solubility at temperatures above 50 °C. He talks about how to improve his investigation. He has three ideas:

- Increase the range by taking readings at 60 °C and 70 °C.
- Repeat the experiment at 10 °C, since the result is anomalous.
- Repeat all readings to make the results more reliable.

> **Questions**
> 1. Define the term *continuous variable*.
> 2. Use the graph to predict the solubility of sodium carbonate at 35 °C.

> **Key points**
> - In a conclusion, describe – and, if possible, explain – any patterns.
> - In an evaluation, describe limitations and suggest improvements.

Review 8.10

1. Choose words from the list to copy and complete the sentences below. [6]

solubility	concentration
pure	dilute
concentrated	purity

   a. A substance that has nothing mixed with it is _____.

   b. _____ describes how much of a substance is in a mixture.

   c. A solution mixed with a large amount of water is _____.

   d. A solution mixed with little water is _____.

   e. The _____ of a solution is a measure of the number of solute particles in a volume of the solution.

   f. The maximum mass of substance that dissolves in 100 g of water is its _____.

2. A student had some liquid salol. She allowed it to cool. Every minute, she measured the temperature of the salol. Her results are in the table below.

Time (min)	Temperature (°C)
0	70
1	56
2	42
3	42
4	42
5	30
6	20

   a. Name the independent variable. [1]

   b. Name the dependent variable. [1]

   c. Plot the points in the table on a graph, and draw a line of best fit. Use the axes below. [3]

   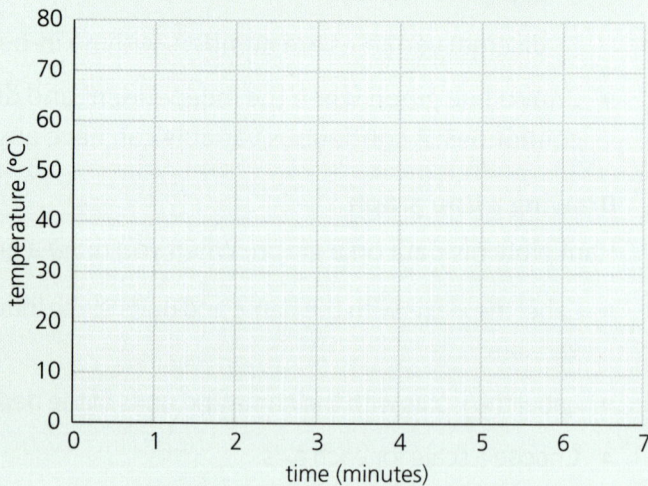

   d. Use your graph to give the melting point of salol. [1]

   e. Describe what happens to the movement and arrangement of the particles when liquid salol freezes. [3]

3. A student heated some liquid water. He recorded the temperature of the water every minute. He plotted his results on a graph.

   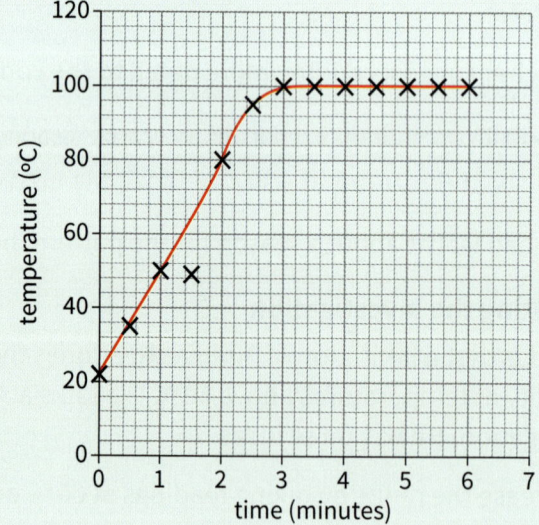

   a. Name the change of state that happens between the third and sixth minutes. [1]

   b. Identify the anomalous result shown on the graph. [1]

   c. Suggest one mistake the student might have made to get this anomalous result. [1]

Pure substances and solutions

4. A student wrote down a question to investigate:

   How does the temperature of water affect the maximum mass of potassium chloride that can dissolve in it?

   a. The student listed some variables in the investigation:

   temperature of water

   volume of water

   mass of potassium chloride that dissolves

   amount of stirring

   i. From the list above, identify the independent variable (the variable the student will change.) [1]

   ii. From the list above, identify the dependent variable. [1]

   iii. From the list above, identify two control variables. [1]

   b. The student made a prediction:

   The hotter the water, the greater the mass of potassium chloride that will dissolve.

   He carried out the investigation, collected results, and plotted the graph below.

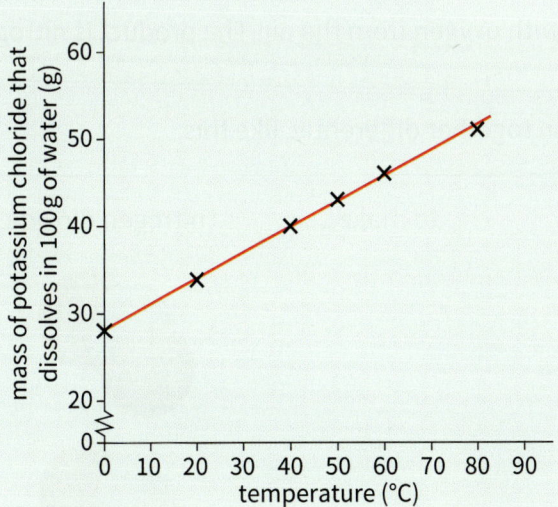

   Explain whether or not the student's prediction was correct. [2]

   c. Explain why the student plotted his results on a line graph, not a bar chart. [1]

5. A student wanted to obtain a chromatogram of black ink. She set up the chromatography apparatus below.

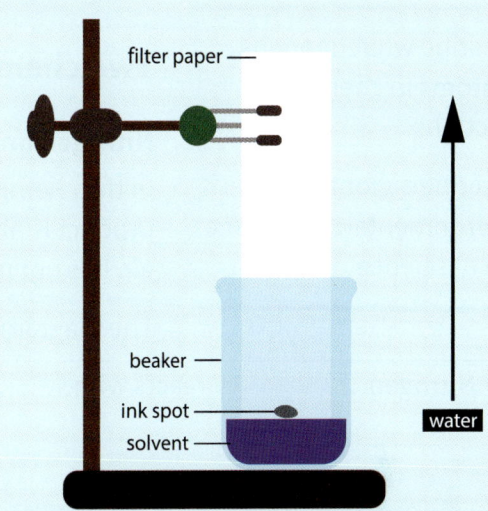

   a. Explain why the student drew a line near the bottom of the chromatography paper with pencil, not ink. [1]

   b. Suggest a solvent the student could use. [1]

   c. The student obtained the chromatogram below.

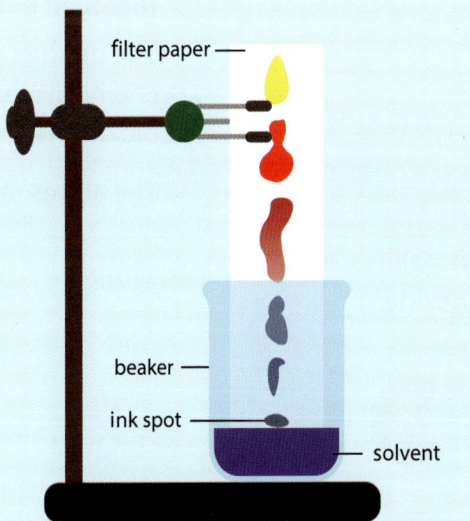

   i. Give the number of dyes that were mixed to make the ink. [1]

   ii. Explain why the dyes in the ink separated. [1]

145

9.1 More chemical reactions

Objectives

- Describe what happens to atoms in chemical reactions
- Describe a global environmental impact of a use of science

At the high temperature of a car engine, nitrogen and oxygen gases react together. What is the environmental impact of this chemical reaction?

Two chemical reactions

Nitrogen and oxygen

In the chemical reaction, nitrogen and oxygen are the reactants. The product is nitrogen monoxide, NO. The nitrogen monoxide gas goes into the air. It is not possible to get the reactants back again.

The diagram below models the reaction. Each sphere represents one atom.

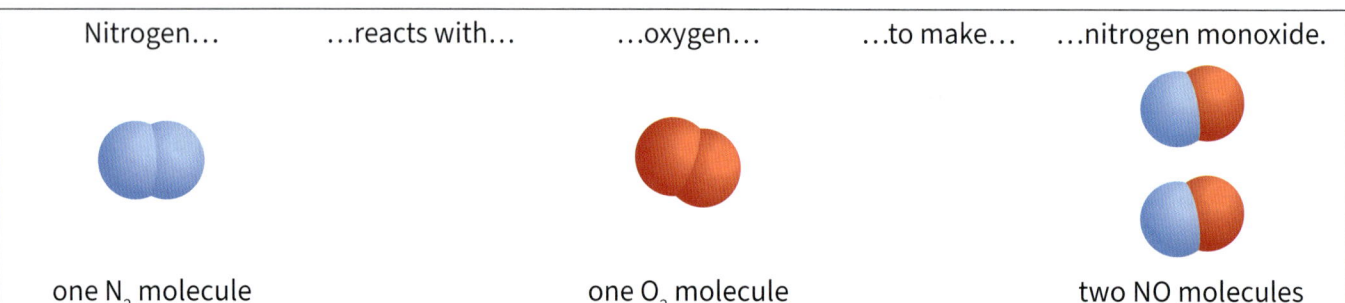

| Nitrogen… | …reacts with… | …oxygen… | …to make… | …nitrogen monoxide. |

one N_2 molecule · one O_2 molecule · two NO molecules

The diagram shows that, in the chemical reaction, the number of nitrogen atoms does not change. The number of oxygen atoms does not change. The atoms rearrange and join together differently.

Nitrogen monoxide and oxygen

The product of the reaction above, nitrogen monoxide, takes part in another chemical reaction. It reacts with oxygen from the air. The product is nitrogen dioxide, NO_2.

The atoms rearrange and join together differently, like this:

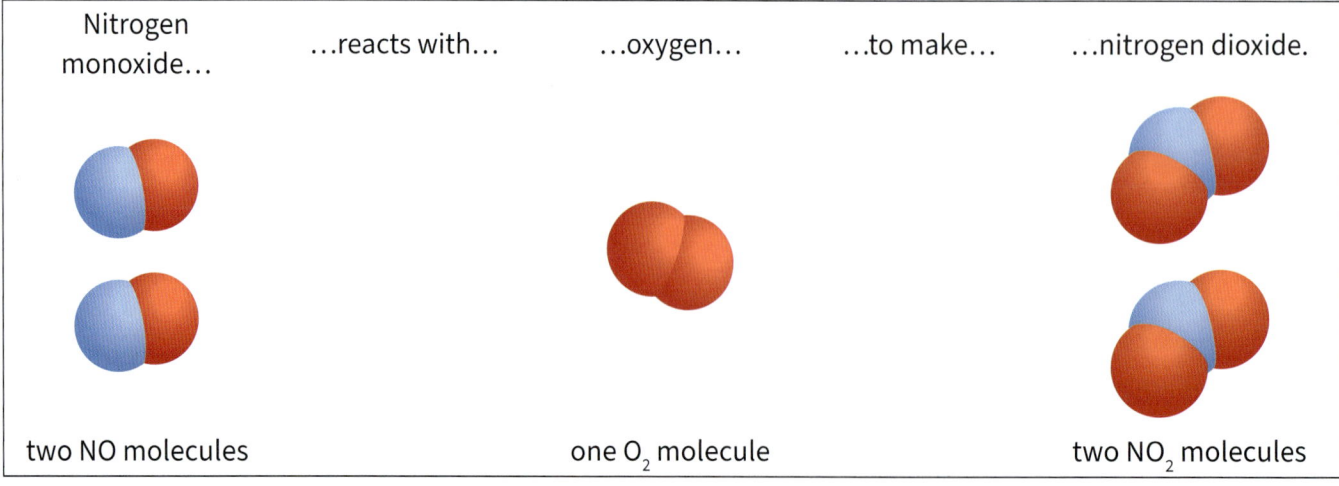

Nitrogen monoxide… …reacts with… …oxygen… …to make… …nitrogen dioxide.

two NO molecules · one O_2 molecule · two NO_2 molecules

As in all chemical reactions, the number of atoms of each element does not change.

146

Chemical reactions 2

🔬 Science in context

The impacts of nitrogen dioxide

Nitrogen dioxide has impacts on the environment, and on health.

Nitrogen dioxide and the environment

In the air, nitrogen dioxide dissolves in rain. This makes acid rain. As you know, acid rain damages some buildings. It harms ecosystems, such as lakes and forests.

Scientists have investigated the effects of acid rain on crops. They found that acid rain damages the waxy coating on the leaves on carrots and other plants. The damage slows their growth.

▲ *Acid rain damages carrot leaves.*

Nitrogen dioxide and health

There may be high levels of nitrogen dioxide gas in the air, both outside and inside:

- Outside, the gas is made in the chemical reactions shown opposite.
- Inside, the gas is made in gas stoves.

Scientists have investigated the effects of nitrogen dioxide on health. If the gas is in the air, then – over time – it damages the lungs. Nitrogen dioxide also makes asthma worse.

▲ *Nitrogen dioxide makes asthma worse.*

Action for health and the environment

There are things you can do to reduce the amount of nitrogen dioxide in the air:

- Outside, perhaps you can walk or cycle instead of going by car.
- Inside, open the windows when a gas stove is burning.

❓ Questions

1. Describe what happens to the atoms in a chemical reaction.
2. Give the total number of atoms in each of these molecules. Use the molecule diagrams on page 146 to help.
   a. A nitrogen molecule, N_2
   b. A nitrogen monoxide molecule, NO
   c. A nitrogen dioxide molecule, NO_2
3. Give the name and formula of the product in each of these chemical reactions:
   a. The reaction of nitrogen with oxygen.
   b. The reaction of nitrogen monoxide with oxygen.
4. Describe two impacts of nitrogen dioxide in the air.

📖 Key points

- In chemical reactions, atoms rearrange and join together differently.
- Nitrogen dioxide harms health and the environment.

9.2

Objective

- Write and interpret word equations

Word equations

A teacher heats some sodium. When the sodium catches fire, she places it in chlorine gas. The sodium continues to burn. White fumes are made. What are the white fumes?

What is a word equation?

The white fumes are sodium chloride, also called salt. Sodium chloride is the product of the chemical reaction of sodium and chlorine. You can represent the reaction in a sentence, like this:

Sodium reacts with chlorine to make sodium chloride.

You can also show the equation with a **word equation**, which is a simple way of representing a chemical reaction.

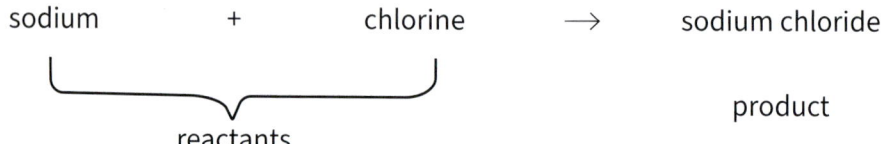

A word equation shows:

- reactants (starting substances) on the left
- products (substances that are made) on the right.

The arrow means 'reacts to make'. In a chemical reaction, the reactants and products are different from each other. So, the arrow in a word equation has a different meaning to the equals sign (=) in a maths equation.

More word equations

You can represent any chemical reaction with a word equation.

A precipitation reaction

Zamira mixes potassium iodide solution with lead nitrate solution. A yellow precipitate of lead iodide forms, along with potassium nitrate solution. The word equation for the precipitation reaction is:

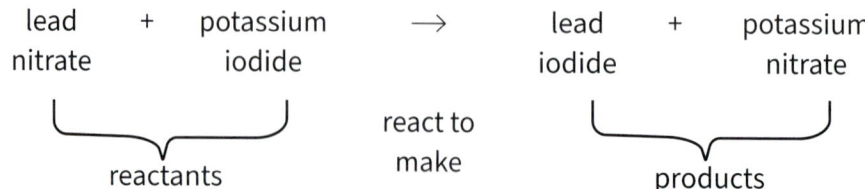

▲ The reaction of potassium iodide and lead nitrate solutions.

148

Chemical reactions 2

A combustion (burning) reaction

Raj burns some magnesium. It reacts with oxygen from the air to make magnesium oxide. The word equation for the chemical reaction is:

magnesium + oxygen ⟶ magnesium oxide

The reactants are magnesium and oxygen. There is one product – magnesium oxide.

A corrosion reaction

When iron corrodes (rusts), it reacts with oxygen and water to make hydrated iron oxide. The word equation for the reaction is:

iron + oxygen + water ⟶ hydrated iron oxide

An acid reaction

One of the acids in acid rain, nitric acid, reacts with calcium carbonate to make three products – calcium nitrate, carbon dioxide, and water. The word equation for the reaction is:

calcium carbonate + nitric acid ⟶ calcium nitrate + carbon dioxide + water

In all word equations, you must write all the reactants and products on the same line. If some of the names are long, you can write them like this:

calcium carbonate + nitric acid ⟶ calcium nitrate + carbon dioxide + water

▲ *Burning magnesium*

▲ *Iron reacts with oxygen and water to make rust.*

Questions

1. Look at the word equation below.

   magnesium + hydrochloric acid ⟶ magnesium chloride + hydrogen

   a. Name the reactants.
   b. Name the products.
   c. Give the meaning of the arrow.

2. Write word equations to show that:

   a. Nitrogen reacts with oxygen to make nitrogen monoxide.
   b. Nitrogen monoxide reacts with oxygen to make nitrogen dioxide.
   c. Iron reacts with sulfur to make iron sulfide.

3. In one chemical reaction, the reactants are zinc and hydrochloric acid. The products are zinc chloride and hydrogen. Write a word equation for the reaction.

Key points

- A word equation is a simple way of representing a chemical reaction.
- In a word equation:
  - the reactants are on the left
  - the products are on the right
  - the arrow means *reacts to make*.

9.3 Energy changes

Objectives

- Define the terms *exothermic* and *endothermic*
- Give examples of exothermic and endothermic changes
- Use temperature change to deduce whether a change is exothermic or endothermic

Kamala cooks over kerosene. Chitra uses charcoal. Grace uses gas. Kerosene, charcoal, and gas are fuels. When a fuel burns, energy is transferred from the fuel. The energy does something useful, such as cook food.

▲ Cooking with charcoal

Energy changes in chemical reactions

All chemical reactions involve energy changes:

- An **exothermic reaction** transfers thermal energy *to* the surroundings. All burning reactions are exothermic. Some neutralisation reactions are exothermic.
- An **endothermic reaction** transfers thermal energy *from* the surroundings. The reaction of citric acid with sodium hydrogencarbonate is endothermic.

You can use temperature changes to work out whether a chemical reaction is exothermic or endothermic.

Hydrochloric acid and sodium hydroxide solution

Tim pours 50 cm³ of hydrochloric acid into a plastic cup. He measures its temperature. He adds 50 cm³ of sodium hydroxide solution. A neutralisation reaction takes place:

hydrochloric acid + sodium hydroxide ⟶ sodium chloride + water

After the reaction, Tim measures the temperature again.

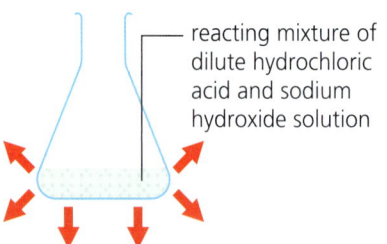

▲ An exothermic reaction transfers energy **to** the surroundings.

	Temperature (°C)
hydrochloric acid, before reaction	21
sodium hydroxide, before reaction	21
reaction mixture, immediately after reaction	52
mixture, 1 hour after reaction	21

The reaction transfers energy from the reactants. During the reaction, the energy heats up the reacting mixture. Then the energy is transferred to the surroundings. The temperature of the surroundings increases as the mixture cools to room temperature.

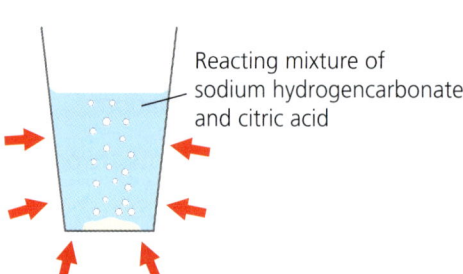

▲ An endothermic reaction transfers energy **from** the surroundings.

Citric acid and sodium hydrogencarbonate solution

Edward pours sodium hydrogencarbonate solution into a plastic cup. He measures its temperature. He adds citric acid powder. The reacting mixture fizzes. After the reaction, Edward measures the temperature again. The temperature is lower.

Chemical reactions 2

During the reaction, the reacting mixture uses thermal energy from the solution. Its temperature decreases. Then energy is transferred from the surroundings. The temperature of the surroundings decreases as the mixture warms to room temperature.

More exothermic and endothermic changes

Changes of state

Changes of state also involve energy changes. Melting and evaporation are endothermic. Condensing and freezing are exothermic.

▲ Imagine holding some ice. The melting ice transfers energy from your hand. The particles use the energy to leave their places in the pattern of solid ice. They start to move around, sliding over each other.

▲ Do you sweat when it's hot? Sweat comes out of pores in your skin. Water from the sweat evaporates. The evaporating sweat transfers energy from your skin. This cools you down.

Dissolving

Dissolving can be exothermic or endothermic. The table gives some examples.

Substance	Does temperature increase or decrease on dissolving?	Exothermic or endothermic?
ammonium chloride	decreases	endothermic
ammonium nitrate	decreases	endothermic
copper sulfate	increases	exothermic
potassium nitrate	increases	exothermic

Science in context

Using an endothermic change

Have you ever used a cold pack on an injury? One type of cold pack involves an endothermic change. An outer bag contains liquid water. An inner bag contains ammonium nitrate powder. When you break the inner bag, the ammonium nitrate dissolves and the mixture cools.

▲ A cold pack

The warm injured body part transfers energy to the cold water. The injury cools and feels better.

Questions

1. Write definitions for *exothermic change* and *endothermic change*.

2. Give examples of:
   a. Two types of change that are always exothermic.
   b. Two types of change that are always endothermic.

3. Sarah has some water. Its temperature is 20 °C. She dissolves a substance in the water, and the temperature changes to 12 °C. State whether the change is exothermic or endothermic.

Key points

- An exothermic change transfers energy to the surroundings. At first, the temperature of the reacting mixture increases.

- An endothermic change transfers energy from the surroundings. At first, the temperature of the reacting mixture decreases.

Thinking and working scientifically 9.4

Objective

- Describe how to plan, carry out, and analyse an investigation

Investigating fuels

Fatima is thinking about fuels. She asks herself some questions.

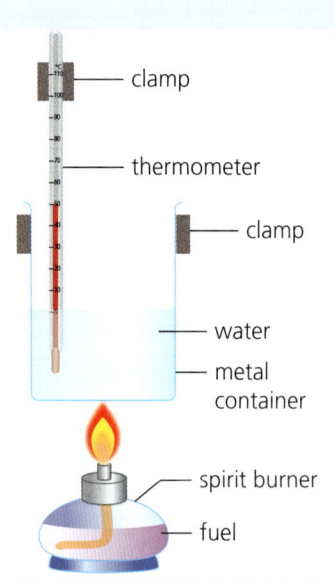

▲ Apparatus to investigate the question: Which fuel makes water hottest?

Planning the investigation

Fatima has three fuels. She decides to investigate the scientific question:

Which fuel makes water hottest?

Her teacher shows her how to set up the apparatus.

Considering variables

Fatima thinks about the variables. She knows that changing one variable will affect another variable. This means she can answer the question by doing a fair test. She lists the variables:

- Independent variable (variable to change) – type of fuel
- Dependent variable (variable to measure) – temperature change of water
- Control variables (variables to keep the same) – mass of fuel burned, volume of water

Carrying out the investigation

Collecting evidence

Fatima measures out $100 \, cm^3$ of water. She uses a thermometer to find its temperature. She lights the burner. The burner heats the water. When 1 g of fuel has burned, she puts out the flame. Then she finds the new temperature.

Fatima decides to take three repeat measurements for each fuel. If the three values are similar, the data are reliable. When she has three values for each fuel, she calculates the mean average.

Presenting evidence

Fatima draws a results table. The independent variable is in the left column. There is a unit in the other column heading. Fatima collects the data opposite.

Fuel	Temperature change (°C)			
	first time	second time	third time	average
ethanol	60	40	40	
propanol	50	53	56	53
butanol	55	57	53	55

The first ethanol measurement is very different from the others. It is anomalous. Fatima repeats the experiment for ethanol one more time. The temperature change is 43 °C, which is close to the second and third values. Fatima calculates the mean value for ethanol:

$$\frac{40 + 40 + 43}{3} = 41\,°C.$$

If Fatima had not made repeat measurements for each fuel, she would not have spotted the anomalous result.

The three results for butanol (55 °C, 57 °C, and 53 °C) are close together. These data are precise.

Analysing the evidence

Making a conclusion

Fatima writes a conclusion:

The temperature change for ethanol was smallest, with a mean value of 41 °C. This means that ethanol is the fuel that transfers least thermal energy. Burning 1 g of propanol increased the water temperature by 53 °C. Burning 1 g of butanol increased the water temperature by 51 °C. This means that propanol and butanol transfer more thermal energy than ethanol.

Fatima does not include a scientific reason in her conclusion, as she has not learnt about this yet.

Suggesting improvements

Fatima notices that some energy from the burning fuel is not transferred to the water. The energy goes into the air. If she repeats the experiment, she will shield the flame so that more of the energy goes into the water.

Questions

1. Describe how to make accurate measurements with a thermometer.
2. Explain why Fatima takes repeat measurements for each fuel.
3. Suggest why Fatima does not investigate the question *Which fuel is best?*

Key points

- An investigation involves planning, carrying out, and analysis.

Extension 9.5

Investigating food energy

Nku has some cashew nuts and some bread. He asks a scientific question:

Which provides more energy – cashew nuts or bread?

Objectives

- Describe how to plan, carry out, and analyse an investigation
- Describe how to measure and calculate the energy transferred when food burns.

Planning the investigation

Making a hypothesis

Nku knows that bread is mainly carbohydrate and that nuts contain oil. He knows that eating oils provides provide more energy than carbohydrates. He uses this scientific knowledge to make a hypothesis. As you know, a hypothesis is a possible explanation. It is based on evidence, and can be tested further. This is his hypothesis:

Nuts are oil rich and bread is carbohydrate rich. Oils provide more energy than carbohydrates. So nuts transfer more energy than bread.

Nku knows he can test his hypothesis. He makes a prediction:

Burning 1g of a nut will heat water up more than burning 1g of bread.

▲ Cashew nuts

▲ Bread

Carrying out the investigation

Equipment

Nku sets up his apparatus. It is like the apparatus for investigating fuels.

Collecting evidence

Nku draws a results table. He decides to make three repeat measurements for each food. If the three values are similar, the data are reliable.

Food	Test number	Water temperature before heating (°C)	Water temperature after heating (°C)	Water temperature change (°C)
1g of a nut	1	21	61	40
	2	21	65	43
	3	22	59	37
1g of bread	1	22	42	20
	2	22	42	20
	3	22	39	17

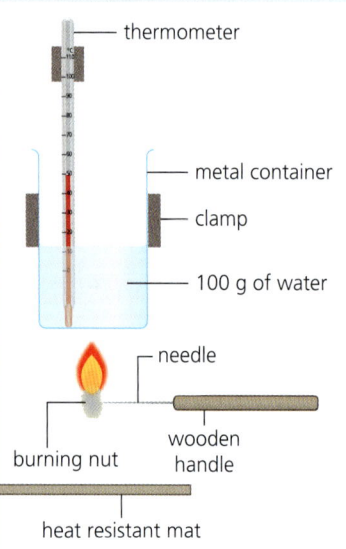

▲ Apparatus to compare the energy transferred when food burns.

Nku calculates the mean water temperature change for each food:

Mean water temperature change for nuts = $\frac{40 + 43 + 37}{3}$ = 40 °C

Mean water temperature change for bread = $\frac{20 + 20 + 17}{3}$ = 19 °C

Analysing the evidence

Writing a conclusion

Nku knows that, the greater the water temperature change, the more energy a burning food transfers.

He writes a conclusion. The conclusion answers his scientific question. He also refers to his prediction and hypothesis.

1 g of burning cashew nut transfers more thermal energy than 1 g of burning bread. My prediction was correct. This supports my hypothesis, that oil-rich cashews transfer more energy.

Evaluating the result

Nku uses a trusted secondary source on the Internet to find this information:

Eating 100 g of cashew nuts transfers 2,200,000 J of energy.

He wants to compare his value to the Internet value.

He calculates the amount of energy his burning nut transfers to the water. He uses this formula: $H = m \times c \times \Delta T$

In the formula:
- H is the energy transferred
- m is the mass of water
- c is the specific heat capacity of water, 4.2 J/g °C
- ΔT is the temperature change of the water.

Nku does the calculation for the nuts.

Energy transferred by 1 g of burning nut
$= m \times c \times \Delta T$
$= 100\,g \times 4.2\,J/g°C \times 40°C$
$= 16,800\,J$

Energy transferred by 100 g of burning nut
$= 16,800\,J \times 100$
$= 1,680,000\,J$

Nku writes an evaluation:

My value is lower than the Internet value. I think that this is because some of the energy from the burning nut was not transferred to the water. It was transferred to the apparatus.

Questions

1. Explain why Nku makes three repeat measurements for each food.
2. A mass of 1 g of burning food heats 100 g of water from 20 °C to 80 °C. Calculate the thermal energy transferred to the water.
3. Explain why the value you calculated in question 2 is likely to be less than the thermal energy transferred by the burning food.

Key points

- An investigation involves planning, carrying out, and analysis.

9.6 Metals and oxygen

Objectives

- Define the term *reactivity*
- Describe the reactivity of metals with oxygen
- Give an example of a chemical reaction that makes a mixture of products

Agir is on a boat. The engine stops working. Agir sets off a flare. He hopes that a sailor in another boat will see the flare and rescue him.

Magnesium and oxygen

The flare includes strontium nitrate, which makes a red flame. The flare also burns magnesium. Magnesium is a metal element. The combustion reaction is vigorous. There is a bright white flame. This word equation summarises the reaction:

magnesium + oxygen $\rightarrow$ magnesium oxide

Magnesium reacts with oxygen even when you do not heat it. In the air, over time, its surface atoms react with oxygen. This makes a thin layer of magnesium oxide. The reaction is a corrosion reaction.

How do other metals react with oxygen?

As you know, metal elements have similar physical properties. They are shiny. They conduct heat and electricity. Metals have patterns in their chemical properties too.

When a metal reacts with oxygen, the product is an oxide. Some metals react with oxygen more easily and vigorously than others.

Sodium burns very vigorously in air. There is not just one product. There is a mixture of products:

- sodium oxide, Na_2O
- sodium peroxide, Na_2O_2

Sodium also reacts with oxygen without heating. As soon as you cut the metal, the shiny surface reacts with oxygen. This makes white sodium oxide. You can see sodium oxide on the uncut surfaces.

Other metals also react with oxygen. The pictures at the top of the next page show some examples.

▲ The magnesium on the left has a layer of magnesium oxide on its surface.

▲ Sodium burns very vigorously in air.

▲ Sodium reacts with oxygen from the air, even without heating.

Chemical reactions 2

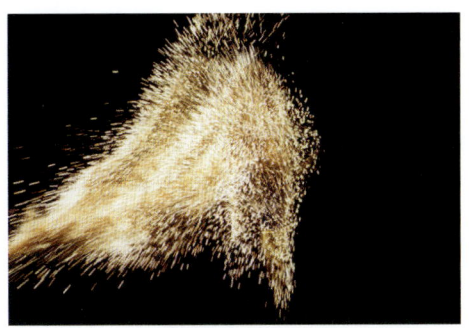

▲ If you sprinkle zinc powder into a Bunsen flame, you see bright white sparks. Zinc oxide forms. Iron powder reacts in a similar way.

▲ Copper does not burn in a Bunsen flame. Instead, it forms black copper oxide on its surface:
copper + oxygen → copper oxide

▲ Gold does not burn. Its surface atoms do not react with oxygen. This explains why gold stays shiny, and why it is used for connectors in audio equipment.

An order of reactivity

The table summarises the reactions of some metals with oxygen.

Metal	Reaction with oxygen	Reactivity
sodium	Burns vigorously. Surface atoms react quickly with oxygen, without heating.	Reactions get less vigorous
magnesium	Burns vigorously. Surface atoms react slowly with oxygen, without heating.	
zinc	Powder burns vigorously. Bigger pieces do not burn easily.	
iron	Powder burns vigorously. Bigger pieces do not burn easily.	
copper	Does not burn. Surface atoms react with oxygen, when heated.	
gold	No reaction.	

The metals get less reactive from top to bottom. Their reactivity decreases.
Reactivity is the tendency of a substance to take part in chemical reactions.

Questions

1. Write the definition of *reactivity*.
2. Name the product of the reaction when zinc burns in air.
3. When sodium burns in air, a mixture of products is formed. Write the names and formulae of the two substances in the mixture.
4. Write a word equation for the reaction of iron and oxygen.
5. List these metals in order of how vigorously they react with oxygen, most reactive first: copper, gold, iron, magnesium, sodium, zinc
6. 'The burning reactions of metals are exothermic.' Give one piece of evidence that supports this statement.

Key points

- Reactivity is the tendency of a substance to take part in chemical reactions.
- Some metals react with oxygen to make oxides.
- Sodium and magnesium react vigorously with oxygen.

9.7 Metals and water

Objective

- Describe the reactivity of metals with water

Do metals react with water? Not the ones we use most often. There would be a problem if metal taps reacted with water, or if metal cars reacted with rain.

▲ A tap is no good if its metal reacts with water.

Which metals react with water?

Calcium

Winton drops a piece of calcium into water. It bubbles quickly. Soon, the bubbling stops. The calcium seems to have disappeared.

The calcium has reacted with the water. The bubbles contained hydrogen gas. This word equation shows the reaction:

calcium + water ⟶ calcium hydroxide + hydrogen

Winton repeats the experiment. He collects the gas, and tests it with a lighted splint. The lighted splint goes out with a squeaky pop. This shows that the gas is hydrogen.

▲ Calcium bubbles quickly in water.

Potassium and sodium

Sodium has an exciting reaction with water. Mr Kibiriti cuts a small piece of sodium. He puts it on the water surface. The sodium zooms around, bubbling vigorously.

The chemical reaction makes sodium hydroxide solution and hydrogen gas:

sodium + water ⟶ sodium hydroxide + hydrogen

Mr Kibiriti adds universal indicator to the new solution. It becomes purple. This shows that the solution is alkaline.

Potassium reacts even more vigorously than sodium. The reacting potassium gets hot enough to set fire to the hydrogen gas:

potassium + water ⟶ potassium hydroxide + hydrogen

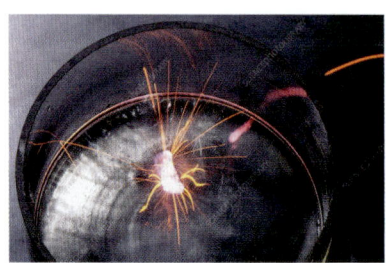

▲ Potassium reacts very vigorously with water.

Magnesium, zinc, and iron

Magnesium reacts slowly with cold water. A few bubbles of hydrogen form on the surface of the metal.

158

Chemical reactions 2

Zinc does not react with cold water. But it does react with steam, which is water in the gas state. The products are zinc oxide and hydrogen.

zinc + water (as steam) ⟶ zinc oxide + hydrogen

Iron does not react with water alone. But, as you know, it corrodes with water and oxygen to make rust (hydrated iron oxide).

Which metals do not react with water?

23 December 2007
Yesterday, Chinese archaeologists raised the wreck of an 800-year-old ship from the depths of the South China Sea. The ship, which sank in heavy storms, was carrying treasure to sell overseas – including gold and silver pots, and 6000 copper coins.

Why did the treasure survive for so long? Because silver, gold, and copper do not react with cold water.

An order of reactivity

The table summarises the reactions of some metals with water.

Metal	Reaction with water	Reactivity
potassium	React vigorously with cold water.	Reactions get less vigorous
sodium		
calcium		
magnesium	Reacts very slowly with cold water.	
zinc	Reacts with water in the gas state (steam).	
iron	Does not react with cold water alone, but reacts with water and oxygen to make rust.	
copper	Do not react with cold water.	
silver		
gold		

All the metals that react with water form two products:

- an oxide or hydroxide
- hydrogen.

The metals that react vigorously with oxygen also react vigorously with water. Is there a similar pattern for the reactions of metals with acids? Turn over to find out.

Questions

1. Name three metals that react vigorously with cold water.
2. Name the products of the reaction of potassium and water.
3. Write a word equation for the reaction of sodium and water.
4. Explain why gold coins stay shiny for many years. In your answer, write about the reactivity of gold with oxygen and water.

Key points

- Some metals react with water to make hydrogen and a metal oxide or hydroxide.
- Potassium, sodium, and calcium react most vigorously with water.
- Gold, silver, and copper do not react with water.

159

9.8 Metals and acids

Objectives
- Describe the reactivity of metals with dilute acids
- Describe how scientists may evaluate an environmental issue

Zookeepers were worried. A hyena was ill. Its foot was swollen. It refused to eat. What was wrong? Vets used X-rays to solve the mystery. The hyena had swallowed zinc coins.

How does zinc react with dilute acids?

Just like you, the hyena has hydrochloric acid in its stomach. The acid reacted with the zinc coins.

zinc + hydrochloric acid ⟶ zinc chloride + hydrogen

Zinc chloride dissolves in water. It mixed with the blood and travelled around the body. This poisoned the hyena.

How do other metals react with acids?

Magnesium and iron

Lola pours dilute hydrochloric acid into a test tube. She adds some magnesium.

The mixture bubbles vigorously. The magnesium ribbon gets smaller, and disappears. A colourless solution remains. There has been a chemical reaction:

magnesium + hydrochloric acid ⟶ magnesium chloride + hydrogen

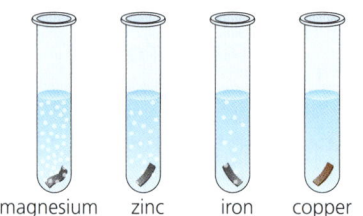

▲ *Magnesium, zinc, and iron react with dilute hydrochloric acid, but copper does not.*

Lola repeats the experiment with different metals. Zinc and iron bubble slowly. Copper does not react with the acid.

There is a pattern in the reactions of metals with dilute acids. All the reactions make hydrogen gas. The other product is a compound that dissolves in water:

magnesium + hydrochloric acid ⟶ magnesium chloride + hydrogen

zinc + hydrochloric acid ⟶ zinc chloride + hydrogen

iron + hydrochloric acid ⟶ iron chloride + hydrogen

Potassium, sodium, and calcium

Potassium, sodium, and calcium react very vigorously with dilute acids. You will not do these reactions in the lab at school.

Copper, silver, and gold

Some metals do not react with dilute acids. Nothing happens if you add gold, silver, or copper to dilute hydrochloric acid.

▲ *Gold does not react with dilute acids.*

Chemical reactions 2

An order of reactivity

The table summarises the reactions of some metals with dilute acids.

Metal	Reaction with dilute acid	Reactivity
potassium	React very vigorously.	Reactions get less vigorous
sodium		
calcium		
magnesium	React steadily.	
zinc		
iron		
copper	Do not react with dilute acids.	
silver		
gold		

Every metal that reacts with dilute acids forms two products:

- A compound that dissolves in water. If the acid is hydrochloric acid, the compound is a chloride.
- Hydrogen gas.

The table above shows that – in their reactions with dilute acids – the reactivity of the metals decreases from potassium to gold.

Science in context

An acidic river

The Rio Tinto in Spain is acidic. Its pH is about 2. The river is red because iron compounds are dissolved in it. Compounds of poisonous metals, such as cadmium, are also dissolved in the river. Few plants and animals can live in its waters.

Scientists have analysed the river water to find out exactly what is in it. They hope the new knowledge will help to solve the environmental problems it causes.

▲ *The Rio Tinto, Spain.*

Key points

- Some metals react with dilute acids to make hydrogen and a compound that dissolves in water.
- Potassium, sodium, and calcium react most vigorously with dilute acids.
- Gold, silver, and copper do not react with dilute acids.

Questions

1. Name three metals that react steadily with dilute acids.
2. Name the products of the reaction of zinc with dilute hydrochloric acid.
3. Write a word equation for the reaction of iron with dilute hydrochloric acid.
4. Describe a test to show that the gas made when magnesium reacts with dilute hydrochloric acid is hydrogen.

9.9 The reactivity series

Objectives

- Define the terms *inert* and *reactivity series*
- Describe how the reactivity series is useful

There used to be copper cables in this trench, next to the railway. One night, thieves stole the copper. Why? What makes copper so valuable?

▲ Thieves stole copper wires from this trench.

Pipes and pans

Copper is valuable because its properties make it useful.

Salsal is a plumber. He uses copper pipes. He chooses copper because of its physical and chemical properties:

- Physical property – copper is malleable. Salsal can bend the pipes to the angles he needs.
- Chemical properties – copper does not react with water flowing through the pipes. It does not react with oxygen from the air.

Kiraz is a cook. She likes copper pans. The properties of copper are perfect for pots:

- Physical property – copper is a good conductor of heat.
- Chemical properties – copper does not react with water, acidic food, or oxygen in the air.

▲ Copper pipes   ▲ A copper pan

Inert metals

Copper reacts with few substances. It is unreactive. Gold is even less reactive. It is **inert**. An inert substance is one that does not take part in chemical reactions.

Inert metal elements include gold, platinum, rhodium, and palladium. Inert non-metal elements include helium, neon, and argon.

Chemical reactions 2

The reactivity series

The patterns of metal reactions with oxygen, water, and acids are similar. The **reactivity series** describes this pattern. It lists the metals in order of how readily they react with other substances.

The metals at the top of the reactivity series have vigorous reactions. Going down the list, the metals get less reactive. The metals at the bottom are inert.

The reactivity series of metals

Reactivity decreases from top to bottom.

potassium
sodium
calcium
magnesium
zinc
iron
copper
silver
gold

Science in context

Using the reactivity series

Ships are made from steel. As you know, steel is an alloy of iron. Iron corrodes when its surface atoms react with oxygen and water.

Paint protects steel ships well, except when scratched. So, some steel boats have pieces of zinc joined to them. Zinc is higher in the reactivity series than iron. The zinc reacts with water and oxygen instead of iron. The zinc is sacrificed to save the iron, so this is called sacrificial protection.

▲ *There are zinc blocks joined to the painted steel boat.*

Questions

1. Write the definitions for *inert* and *reactivity series*.
2. Look at the reactivity series.
   a. Name three metals that are more reactive than magnesium.
   b. Name three metals that are less reactive than magnesium.
3. Predict the products of the reaction of calcium with dilute hydrochloric acid.
4. Zinc protects steel ships by sacrificial protection. Steel is mainly iron. Explain why magnesium also protects ships in this way.

Key points

- An inert substance does not take part in chemical reactions.
- The reactivity series lists the metals in order of how readily they react with other substances.

Thinking and working scientifically

9.10

Objective

- Describe how to plan, carry out, and analyse an investigation

▲ Health hazard

▲ Environmental hazard

▲ Harmful

Lead in the reactivity series

Manbir works in a hospital. He takes X-ray images. Inside his apron are sheets of a metal element, lead. The lead stops X-rays damaging his internal organs. Lead is also used as a roofing material, since it is waterproof and easy to bend.

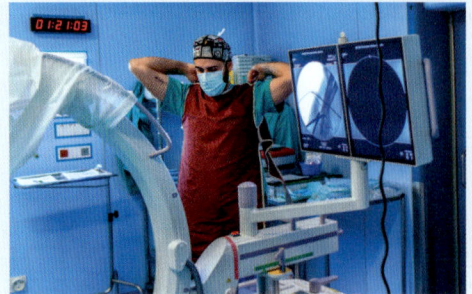

Planning an investigation

Deepa wants to find the position of lead in the reactivity series. She plans how to investigate its chemical properties.

Risk assessment

A container of lead has the hazard symbols opposite. The element is poisonous, and can be stored in the body.

Deepa decides to reduce the risk of injury or damage by not using lead herself. Instead, she will obtain evidence from secondary sources.

Hypothesis

Deepa watches a video of the reaction of lead with hydrochloric acid. Bubbles form on the surface of the metal. The bubbling is slower than in the reaction of zinc with acid, which Deepa did last week.

Deepa makes a hypothesis. As you know, a hypothesis is a possible explanation. It is based on evidence, and can be tested further:

Lead is between zinc and copper in the reactivity series. I think this because lead and acid bubble more slowly than zinc and acid. And copper does not react with acid.

Variables

Deepa thinks about the variables. She decides to do a fair test, since changing one variable will affect another variable:

- independent variable (variable to change) – metal
- dependent variable (variable to measure or observe) – how much it bubbles
- control variables (variables to keep constant) – concentration of acid, size of metal pieces, temperature of acid.

Carrying out the investigation

Deepa watches the video of the reaction of lead with hydrochloric acid again. She then does the same experiment in the lab with copper, iron, and zinc. She tries to keep the control variables the same as in the video.

Deepa records her observations in a table.

Metal	Observations	Source
lead	steady bubbling	Internet (university chemistry department)
copper	no reaction	my own experiment
iron	steady bubbling – faster than lead	my own experiment
zinc	fast bubbling	my own experiment

Analysing the evidence

Making a conclusion

Deepa looks at her results. She lists the metals in order of how vigorously they react with dilute acid from most reactive to least reactive.

zinc, iron, lead, copper

She writes a conclusion:

Lead is between iron and copper in the reactivity series. This supports my hypothesis.

Limitations

Deepa does not know the exact size of the lead in the video. She does not know the acid temperature. This means that she cannot be sure that her test is fair. She also says that it is difficult to compare the bubbling.

Suggesting improvements

Deepa suggests doing two more experiments:

- reacting the metals with oxygen
- reacting the metals with water.

If the reactivity of lead is between iron and copper with oxygen and with water, Deepa can be more certain that her conclusion is correct.

Questions

1. Write the definitions of *independent variable*, *dependent variable*, and *control variable*.
2. Explain why Deepa collects some of her evidence from a secondary source.
3. Predict what Deepa might observe if she adds cold water to a piece of lead. Give a reason for your prediction.

Key points

- An investigation involves planning, carrying out, and analysis.

Review 9.11

1. Name the reactants and products in the reactions shown by the word equations below.

    a. sodium + iodine ⟶ sodium iodide [2]

    b. carbon + oxygen ⟶ carbon dioxide [2]

    c. magnesium + hydrochloric acid ⟶ magnesium chloride + hydrogen [2]

    d. lithium + water ⟶ lithium hydroxide + hydrogen [2]

2. Copy and complete the word equations below.

    a. sodium + chlorine ⟶ _____ [1]

    b. zinc + _____ ⟶ zinc oxide [1]

    c. _____ + sulfur ⟶ iron sulfide [1]

    d. iron + oxygen ⟶ _____ [1]

    e. sulfur + _____ ⟶ sulfur dioxide [1]

3. Write word equations to summarise the reactions below.

    a. Burning sodium in oxygen to make sodium oxide. [3]

    b. Heating calcium in air to make calcium oxide. [3]

    c. Reacting sodium chloride and silver nitrate solutions to make silver chloride and sodium nitrate. [3]

    d. Reacting zinc with hydrochloric acid to make zinc chloride and hydrogen. [3]

4. Tick the table to show which changes are exothermic and which changes are endothermic. [5]

Type of change	Is the change exothermic?	Is the change endothermic?
combustion (burning)		
evaporation		
melting		
freezing		
condensing		

5. The table gives the temperature changes when different substances dissolve in water. The same mass of solid was used in each test, and the same volume of water.

Substance	Temperature change (°C)
ammonium chloride	−5
ammonium nitrate	−6
concentrated sulfuric acid	+12
copper sulfate	+5
potassium nitrate	−9

   a. Identify the substances that dissolve exothermically in water. [1]

   b. Identify the substance that transfers the most energy from the surroundings when it dissolves. [1]

6. Mr Chinyanganya sets up the apparatus below.

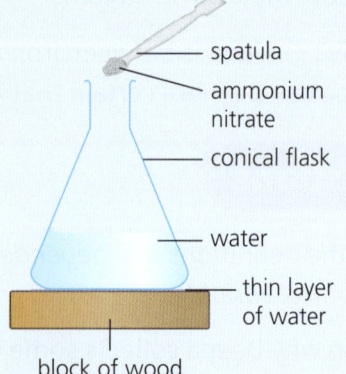

He adds ammonium nitrate to the water, and stirs until it dissolves.

a. The water between the flask and the wooden block freezes. Explain why. [2]

b. State whether ammonium nitrate dissolves exothermically or endothermically. Explain how you know. [2]

7. Ashok compares the heat released in different neutralisation reactions. He uses the apparatus below.

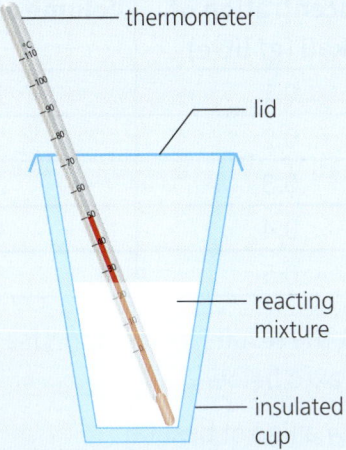

Ashok repeats his investigation three times. The table summarises his results.

Reactants	Temperature change (°C)			
	first time	second time	third time	average
nitric acid and sodium hydroxide	50	32	32	32
hydrochloric acid and potassium hydroxide	34	33	35	34
sulfuric acid and sodium hydroxide	30	34	32	32
sulfuric acid and potassium hydroxide	33	33	33	33

a. Suggest why Ashok repeated his investigation three times. [1]

b. Ashok ignored his first result for the first pair of reagents. Suggest why. [1]

c. Use the data to write a conclusion for the investigation. [1]

8. Some metals react with oxygen. Write word equations for the reactions of the metals below with oxygen.

a. magnesium [1]

b. zinc [1]

c. potassium [1]

9. The list below shows part of the reactivity series of metals.

sodium
magnesium
zinc
iron
lead
silver
gold
platinum

a. Name the most reactive metal in the reactivity series shown above. [1]

b. Name one metal that is more reactive than lead. [1]

c. Predict what you would observe if you added platinum to dilute hydrochloric acid. [1]

d. Predict what you would observe if you left a piece of platinum in a beaker of water for a week. Explain why you made this prediction. [2]

e. From the list, suggest a metal that could be attached to an iron ship so that the metal would corrode instead of the iron ship. [1]

Stage 8 Review

1. The diagram shows the sub-atomic particles in an atom of an element.

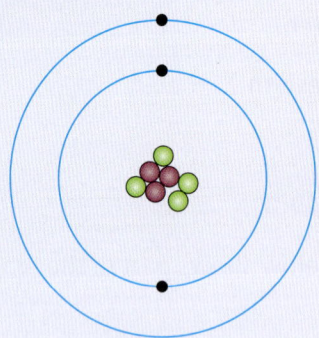

   a. Give the number of sub-atomic particles in the nucleus of the atom. [1]
   b. Name the two types of particles found in the nucleus. Give the relative charge and mass of each type of particle. [2]

2. A student investigates the question below:

   How does the volume of acid needed to neutralise an alkali depend on the concentration of the alkali?

   The student writes down what she plans to do:
   - Measure out 50 cm³ of alkaline solution.
   - Add universal indicator.
   - Add acid until the indicator shows the solution is neutral.
   - Write down the volume of acid used.
   - Repeat with alkaline solution of different concentrations.

   a. The student uses a measuring cylinder, not a beaker, to measure the volume of acid added. Suggest why. [1]
   b. Copy and complete the table. Use the phrases below.

      volume of acid
      concentration of alkali
      type of indicator
      concentration of acid
      volume of alkali

Independent variable (variable to change)	
Dependent variable	
Control variables	1. 2. 3.

   [5]

   c. The student's results are in the table.

Concentration of alkali (g/ litre)	Volume of acid (cm³)
0.1	10
0.2	20
0.3	30
0.4	40
0.5	50

   i. Plot the results on a graph. Use a copy of the axes below. [3]
   ii. Draw a line of best fit. [1]
   iii. Write a conclusion for the investigation. [2]

   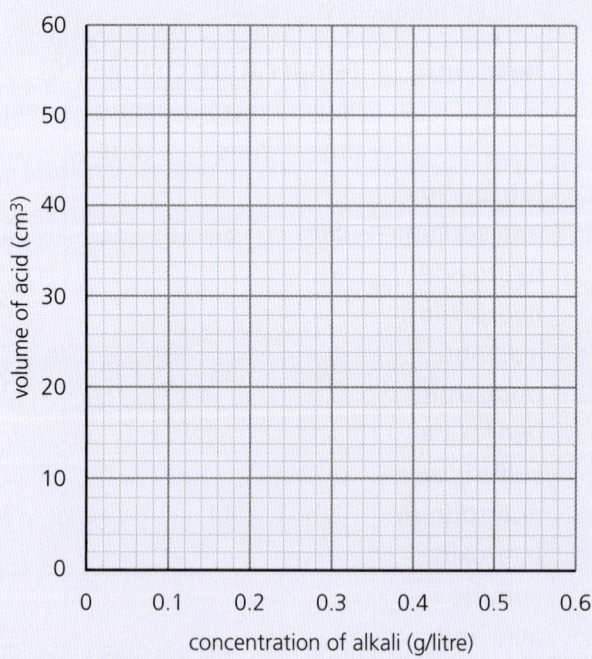

3. Zoza investigates a neutralisation reaction. She adds 50 cm³ potassium hydroxide solution to 50 cm³ of nitric acid.
   Her results are in the table above right.

Stage 8 Review

	Temperature (°C)
potassium hydroxide, before reaction	23
nitric acid, before reaction	23
reaction mixture, immediately after reaction	56
reaction mixture, two hours after reaction	23

a. Use data from the table to calculate the temperature change during the reaction. [1]

b. Explain how the data show that the neutralisation reaction is exothermic. [2]

c. Name three other types of change that are exothermic. [3]

4. Marcus investigates the energy given out by four fuels. He sets up the apparatus below. He measures the temperature change of the water when he burns 1 g of each fuel.

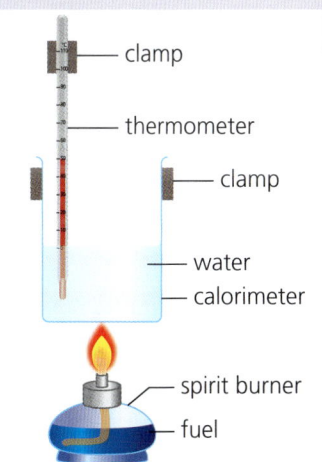

a. Name the independent variable (the variable he changes) and the dependent variable. [2]

b. Name two variables Marcus must control in the investigation. [2]

Fuel	Temperature before heating (°C)	Temperature after heating (°C)	Temperature change (°C)
butanol	20	73	53
pentanol	21	70	
hexanol	22		50
heptanol	21	73	52

c. Marcus obtains the results in the table.

  i. Calculate the two missing values in the table. [2]

  ii. Name the fuel that transferred most energy to the water. [1]

d. Marcus writes a conclusion for his investigation.

The temperature change was similar for all the fuels. I think 1 g of all the fuels transfer similar amounts of energy when they burn.

Do you agree with the conclusion? Explain why, or why not. [1]

e. Marcus thinks that some of the energy from the burning fuels was not transferred to the water. Suggest two other things that the energy might have been transferred to. [2]

5. Copy and complete the word equations below.

a. _____ + bromine ⟶ iron bromide [1]

b. copper + _____ ⟶ copper oxide [1]

c. magnesium + chlorine ⟶ _____ [1]

d. nitrogen + _____ ⟶ nitrogen dioxide [1]

6. The list shows part of the reactivity series:

> potassium
> sodium
> calcium
> iron
> silver
> gold

a. A teacher adds a small piece of sodium to water.

  i. Describe what the teacher would see. [1]

  ii. Name the two products of the reaction. [2]

  iii. Name one metal that reacts more vigorously with water than sodium. [1]

  iv. Name two metals that do not react with water. [1]

b. Copy and complete the word equations below.

  i. calcium + water ⟶ [1]

  ii. magnesium + oxygen ⟶ [1]

  iii. zinc + hydrochloric acid ⟶ [1]

169

10.1 Proton number and the periodic table

Objectives

- Define the term *proton number* and *atomic number*
- Describe how proton number gives an element its position in the periodic table

MRI scans unlock some of the brain's deepest secrets. It is thanks to knowledge about atoms – and what happens inside them – that scientists have developed techniques like this.

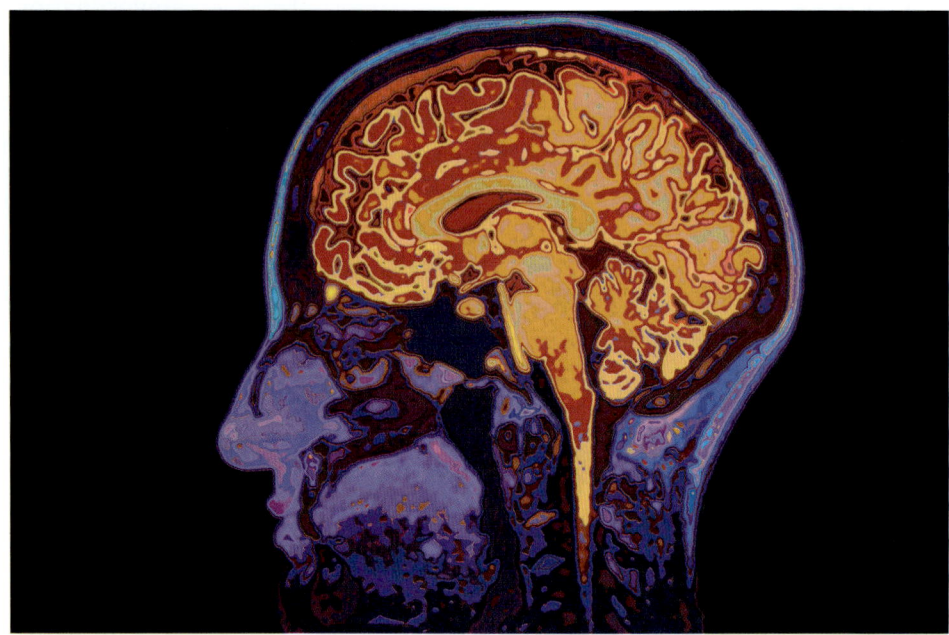

What is proton number?

As you know, an atom has a nucleus. The nucleus is made up of two types of sub-atomic particle:

- positively charged protons
- neutrons, with no charge.

Each element has a different number of protons in its atoms. A helium atom has two protons. A lithium atom has three protons. A gold atom has 79 protons. The number of protons in an atom of an element is its **proton number**. Proton number is also called **atomic number**.

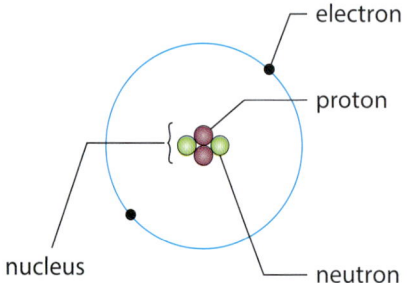

▲ A helium atom has two protons. The proton number of helium is 2.

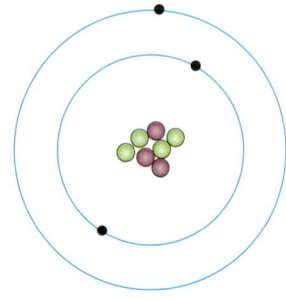

▲ A lithium atom has three protons. The proton number of lithium is 3.

170

Structure, bonding, and properties

How is proton number linked to the periodic table?

In the periodic table, the elements are arranged in order of proton number.

1	2	3	4	5	6	7	8	9	10	11	12	13	14	15	16	17	18
1 hydrogen																	2 helium
3 lithium	4 beryllium											5 boron	6 carbon	7 nitrogen	8 oxygen	9 fluorine	10 neon
11 sodium	12 magnesium											13 aluminium	14 silicon	15 phosphorus	16 sulfur	17 chlorine	18 argon
19 potassium	20 calcium	21 scandium	22 titanium	23 vanadium	24 chromium	25 manganese	26 iron	27 cobalt	28 nickel	29 copper	30 zinc	31 gallium	32 germanium	33 arsenic	34 selenium	35 bromine	36 krypton
37 rubidium	38 strontium	39 yttrium	40 zirconium	41 niobium	42 molybdenum	43 technetium	44 ruthenium	45 rhodium	46 palladium	47 silver	48 cadmium	49 indium	50 tin	51 antimony	52 tellurium	53 iodine	54 xenon
55 caesium	56 barium	57–71 lanthanoids	72 hafnium	73 tantalum	74 tungsten	75 rhenium	76 osmium	77 iridium	78 platinum	79 gold	80 mercury	81 thallium	82 lead	83 bismuth	84 polonium	85 astatine	86 radon
87 francium	88 radium	89–103 actinoids	104 rutherfordium	105 dubnium	106 seaborgium	107 bohrium	108 hassium	109 meitnerium	110 darmstadtium	111 roentgenium	112 copernicium	113 nihonium	114 flerovium	115 moscovium	116 livermorium	117 tennessine	118 oganesson

57 lanthanum	58 cerium	59 praseodymium	60 neodymium	61 promethium	62 samarium	63 europium	64 gadolinium	65 terbium	66 dysprosium	67 holmium	68 erbium	69 thulium	70 ytterbium	71 lutetium
89 actinium	90 thorium	91 protactinium	92 uranium	93 neptunium	94 plutonium	95 americium	96 curium	97 berkelium	98 californium	99 einsteinium	100 fermium	101 mendelevium	102 nobelium	103 lawrencium

▲ This periodic table shows the proton number of each element.

Questions

1. Write the definition for *proton number*.
2. A boron atom has 5 protons and 6 neutrons. Draw and label a diagram of the nucleus of the atom.
3. Use the periodic table to give the proton numbers of these elements:
   a. oxygen, O
   b. chlorine, Cl
   c. silver, Ag
4. Give the names and chemical symbols of the elements with these proton numbers:
   a. 23
   b. 18
   c. 26

Key points

- The proton number of an element is the number of protons in its atoms.
- In the periodic table, the elements are arranged in proton number order.

10.2 Electrons in atoms

Silver and sodium are both metal elements. Many of their physical properties are the same. But their chemical properties are different. Why?

Objectives

- Define the term *electron configuration*
- Draw the electron configurations of 20 elements

▲ Silver (left) and sodium (right) have similar physical properties. They are shiny when freshly cut. They conduct electricity.

▲ Silver and sodium have different chemical properties. Sodium reacts vigorously with water (above). Silver does not react with water.

How many electrons?

As you know, atoms are made up of three types of sub-atomic particle:

- protons and neutrons in the nucleus
- negatively charged electrons, which orbit outside the nucleus.

A neutral atom has the same number of protons and electrons. For example, a hydrogen atom has one proton and one electron. A sodium atom has 11 protons and 11 electrons. A silver atom has 47 protons and 47 electrons.

As you know, every element has a different number of protons in its atoms. This means that every element also has a different number of electrons.

The number and arrangement of electrons gives an element its chemical properties. Sodium and silver have different chemical properties because their atoms have different numbers of electrons.

How are electrons arranged in atoms?

The electrons in an atom are arranged in shells. A sodium atom has:

- two electrons in the first shell, nearest the nucleus
- eight electrons in the second shell
- one electron in the outer shell, furthest from the nucleus.

In atoms of all elements, each electron shell holds a maximum number of electrons:

- The first shell holds up to 2 electrons.
- The second shell holds up to 8 electrons.
- The third shell holds up to 18 electrons.

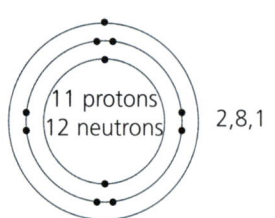

▲ The electron configuration of a sodium atom. Each circle represents one electron.

Structure, bonding, and properties

The arrangement of the electrons in an atom is its **electron configuration**. The diagram below shows the electron configurations of the first 18 elements of the periodic table. The elements are arranged as they are in the periodic table.

Can you see a pattern in the electron structures? Atoms of elements in the left column (Group 1 of the periodic table) have one electron in their outer shell. Atoms of elements in the next column (Group 2) have two electrons in their outer shell. Atoms of all elements that are in the same column of the periodic table have the same number of outer shell electrons.

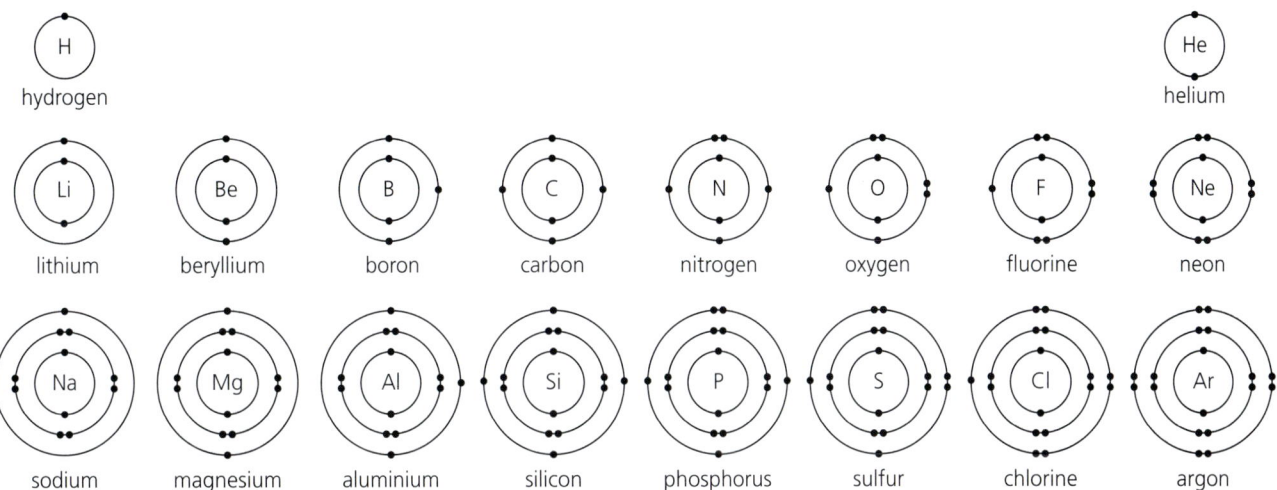

▲ *Electron configuration of the first 18 elements.*

This chapter shows how electron configurations give an element its structure and – as a result – its physical properties. The next chapter shows how electron configurations explain patterns in elements' chemical properties.

Questions

1. Write the definition for *electron configuration*.
2. Give the number of electrons in atoms of:
   a. nitrogen
   b. fluorine
   c. aluminium
3. Draw the electron configurations of atoms of:
   a. lithium
   b. sodium
   c. potassium
4. Look at the electron configurations you drew to answer question 3. Describe how they are similar and how they are different.

Key points

- The electron configuration of an atom describes how its electrons are arranged.
- Each electron shell has a maximum number of electrons.

173

10.3 Making ions

The goat is licking salt, sodium chloride. Why?

Objectives

- Define the term *ion*
- Describe how ions are made
- Explain why atoms form ions
- Write formulae for simple ions

The goat gets sodium ions from salt. Like other animals, the goat needs sodium ions to makes its heart and nerves work.

What are ions?

An **ion** is a particle with a positive or negative charge. An ion forms when an atom gains or loses electrons. Electrons are negatively charged, so:

- If an atom gains one or more electrons, it becomes a negatively charged ion.
- If an atom loses one or more electrons, it becomes a positively charged ion.

A sodium ion forms when a sodium atom gives one electron to a non-metal atom. This happens in a chemical reaction.

Before the reaction:

- The sodium atom has 11 positive protons and 11 negative electrons. It has no net charge.
- The chlorine atom has 17 positive protons and 17 negative electrons. It has no net charge.

In the reaction, one electron moves from the sodium atom to the chlorine atom.

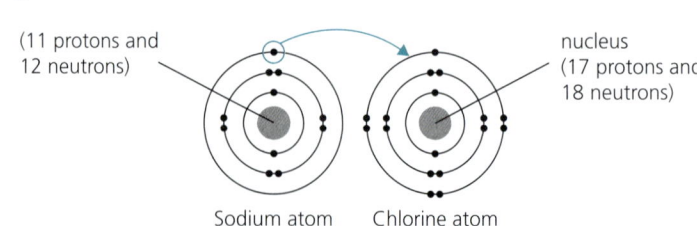

This makes a sodium ion and a chloride ion.

- The sodium ion has 11 positive protons and 10 negative electrons. Its charge is +1.
- The chloride ion has 17 positive protons and 18 negative electrons. Its charge is −1.

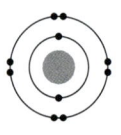 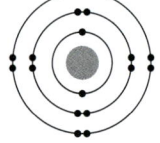

Sodium ion   Chloride ion

Structure, bonding, and properties

Thinking and working scientifically

Formulae for ions

Every ion has its own chemical formula. The formula gives the chemical symbol of the element and the charge on the ion. The formula of a sodium ion is Na^+. The formula of a chloride ion is Cl^-. When you write the chemical formula of an ion:

- Write the + or − sign on the right of the chemical symbol.
- Write the + or − sign above the line.

Why do atoms form ions?

Stable atoms

The element argon makes up 1% of the air. Argon does not take part in chemical reactions. It is inert. Argon is inert because of its electron configuration. Its atoms have eight electrons in the outer shell. The outer electron shell is full. This makes the atom stable.

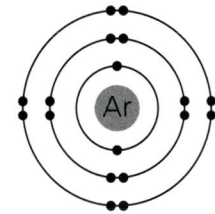

▲ *The electron configuration of an argon atom.*

Making stable ions

Any atom with a full outer electron shell is stable. Ions form in order to achieve this stable structure. As you can see in the diagrams below left:

- A sodium ion has eight electrons in its outer shell. Its outer shell is full. The ion is stable.
- A chloride ion has eight electrons in its outer shell. Its outer shell is full. The ion is stable.

The compound sodium chloride is made up of stable ions. This explains why sodium chloride takes part in few chemical reactions.

Questions

1. Write the definition for *ion*.
2. Most ions have eight electrons in the outer shell. Explain why.
3. Write the chemical formula for each ion below. Use the periodic table to find the chemical symbols of the elements:
   a. A potassium ion, with a charge of +1.
   b. A magnesium ion, with a charge of +2.
   c. A bromide ion, with a charge of −1.
4. The electronic configuration of a fluorine atom is shown. Predict the charge on its ion. Explain your prediction.

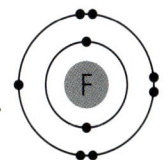

Key points

- An ion is an atom that has gained one or more electrons to be negatively charged, or lost one or more electrons to be positively charged.
- Atoms form ions to gain a stable electron configuration.
- The chemical formula of an ion shows the chemical symbol of the element and the charge on the ion.

10.4 Inside ionic compounds

The photo shows a large crystal of salt, sodium chloride (NaCl). The crystal is made up of sodium ions and chloride ions. What holds the ions together?

Objectives

- Define the terms *ionic bonding, ionic compound,* and *giant ionic structure*
- Explain how the giant ionic structure model explains the properties of ionic compounds
- Discuss the strengths and weaknesses of a physical model

What is ionic bonding?

The salt crystal is made up of millions of sodium ions and millions of chloride ions. Electrostatic attraction between the positive ions and negative ions holds the crystal together. This is **ionic bonding**.

Ionic bonds act in all directions. In the solid state, they hold the positive and negative ions in a three-dimensional pattern. The pattern is called a **giant ionic structure**.

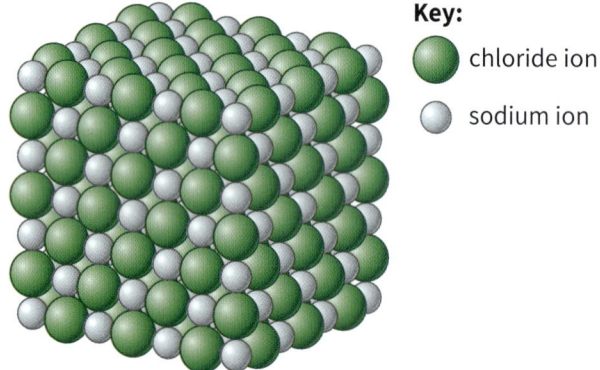

▲ *In sodium chloride, ionic bonds hold the ions together in a giant structure.*

Because it is made up of ions, sodium chloride is an **ionic compound**. Most compounds made up of a metal and a non-metal are ionic.

▲ *Calcium oxide (CaO), nickel chloride ($NiCl_2$), and cobalt chloride ($CoCl_2$) are ionic compounds.*

Ionic bonding and physical properties

As you know, a model is an idea that explains observations and helps in making predictions. The idea of a giant ionic structure is a model. The model explains the physical properties of ionic compounds:

- Ionic compounds have high melting points. This is because the electrostatic attraction between oppositely charged ions is strong.
- Ionic compounds are brittle. If you drop a crystal of an ionic compound, it breaks between one row of ions and another. The broken pieces have straight edges.

Thinking and working scientifically

Modelling ionic bonding

Barney and Sarah use grapes to make a model of a substance with a giant ionic structure, sodium chloride. The model is a physical model, not just an idea.

The model has strengths and limitations. Its main strength is that it shows the positions of the oppositely charged ions. One limitation is that it does not show that the ions vibrate on the spot.

▲ A model of a giant ionic structure

It does not show how the electrostatic attraction acts in all directions. It also does not explain why ionic compounds are brittle.

Questions

1. Write definitions for the terms *ionic bonding*, *ionic compound*, and *giant ionic structure*.
2. Name the type of attraction that holds positive and negative ions together in an ionic compound.
3. Explain these physical properties of an ionic compound:
   a. It has a high melting point.
   b. It is brittle.
4. Describe two strengths and two limitations of the grape model of sodium chloride.

Key points

- An ionic compound is made of positive and negative ions.
- Ionic bonding is the electrostatic attraction between positive and negative ions.
- A giant ionic structure is the three-dimensional pattern of oppositely charged ions.
- Ionic compounds are brittle and have high melting points.

10.5 Covalent bonding

Objectives

- Define the term *covalent bond*
- Draw dot-and-cross diagrams to show shared electron pairs in simple molecules

What comes out of the gills of a fish?

When fish digest food, one of the waste products is ammonia. The ammonia leaves the fish as a gas, through its gills. Ammonia has a bad smell.

Making covalent bonds

Inside ammonia

Ammonia is a compound. Its formula is NH_3. This shows that it is made up of atoms of two elements, nitrogen and hydrogen. There are three hydrogen atoms for every one nitrogen atom.

Ammonia exists as molecules. As you know, a molecule is a particle made up of two or more atoms, strongly joined together. In ammonia, each molecule has one nitrogen atom joined to three hydrogen atoms. The atoms are held together by covalent bonds. A **covalent bond** is a shared pair of electrons that joins two atoms together.

Why form covalent bonds?

Here are the electron configurations of nitrogen and hydrogen atoms:

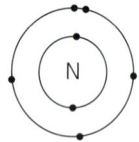
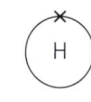

▲ *Nitrogen*   ▲ *Hydrogen*

The nitrogen atom has five electrons in its outer shell. On its own, the atom is not stable. It needs three more electrons to fill its outer shell. It will then have a stable electron configuration, with eight outer electrons.

The hydrogen atom has one electron in its outer shell. On its own, the atom is not stable. It needs one more electron to fill its electron shell. The atom will then have a stable electron configuration.

In ammonia, nitrogen and hydrogen atoms achieve full outer shells by sharing electrons. Each shared pair of electrons is one covalent bond.

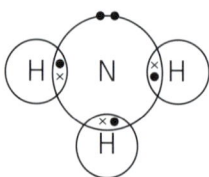

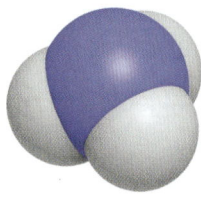

▲ A model of an ammonia molecule. This model shows the atom arrangement. It does not show the shared electron pairs in covalent bonds.

▲ In this diagram, dots show electrons from the outer shell of the nitrogen atom. Crosses show electrons from hydrogen atoms. Each pair of electrons is a covalent bond. All the electrons are the same.

Structure, bonding, and properties

Which substances have covalent bonds?

Compounds of non-metals

Ammonia is a compound of two non-metals. Most other compounds of non-metals exist as molecules. In each molecule, every atom has a share in a full outer shell of electrons. The diagrams below show the outer electron shells only.

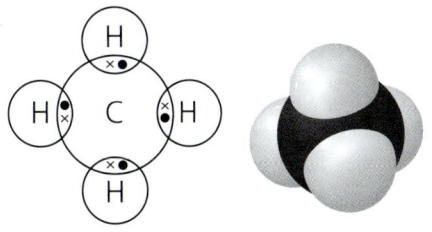

▲ methane, CH_4

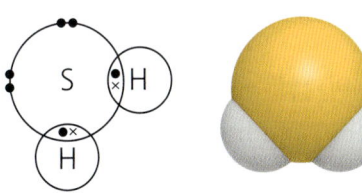

▲ hydrogen sulfide, H_2S

Non-metal elements

The non-metal elements helium, neon, argon, krypton, and xenon exist as single atoms. Most other non-metal elements exist as molecules, with the atoms joined by covalent bonds. In each molecule, every atom has a share in a full outer shell of electrons. The diagrams below show the outer electron shells only.

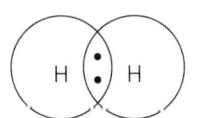

▲ Hydrogen, H_2. Each atom contributes one electron to the shared pair in the covalent bond.

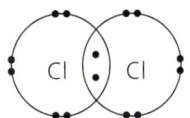

▲ Chlorine, Cl_2. Each atom contributes one electron to the shared pair in the covalent bond.

Ionic or covalent?

Some substances have ionic bonds and other substances have covalent bonds:

- Ionic bonds form in compounds of a metal with a non-metal.
- Covalent bonds form in compounds of non-metals, and in non-metal elements.

Questions

1. Write the definition for a *covalent bond*.
2. Draw a dot and cross diagram of methane, CH_4. Add labels to the diagram to show how each atom has a share in a full outer shell of electrons.
3. A fluorine atom has 7 electrons in its outer shell. Draw a diagram to show the bonding in a fluorine molecule, F_2. Show the outer shells only.
4. Draw a dot-and-cross diagram to show the bonding in hydrogen chloride, HCl. Show the outer shells only.

Key points

- A covalent bond is a shared pair of electrons that joins two atoms together.

10.6 Covalent structures

Objectives

- Define the term *giant covalent structure*
- Explain the properties of substances with simple molecules and giant covalent structures

Carbon and nitrogen are non-metal elements. A carbon atom has six electrons. A nitrogen atom has seven electrons.

At 20 °C, nitrogen is a colourless gas. At the same temperature, carbon is in the solid state. One type of carbon, diamond, is sparkly and hard. Why are the properties of nitrogen and diamond so different?

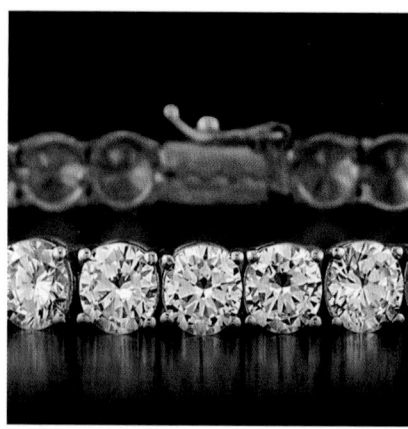

Simple molecule or giant covalent structure?

Nitrogen exists as molecules. But diamond has a giant covalent structure. The different structures explain the different properties.

Simple molecule

Each nitrogen molecule has two atoms. The two atoms are joined by three shared electron pairs. This makes a strong covalent bond.

Nitrogen molecules are attracted to each other only weakly, so little energy is needed to disrupt the pattern of molecules when solid nitrogen melts. This gives nitrogen its low melting point. Other substances that exist as simple molecules – such as oxygen and methane – also have low melting points.

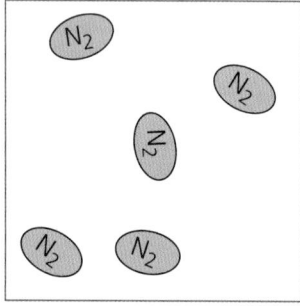

◀ Inside each nitrogen molecule, there is a strong covalent bond. The simple molecules are attracted to each other only weakly.

Giant structure

As you know, diamond is a type of carbon. Each carbon atom makes strong covalent bonds with four other carbon atoms. The pattern is repeated many, many times to make a giant covalent structure. A **giant covalent structure** is a three-dimensional network of atoms that are joined together by strong covalent bonds.

The structure and bonding of diamond explain its hardness and its high melting point, of 3550 °C; large amounts of energy are needed to break its covalent bonds when it melts. Other substances that exist as giant covalent structures, such as silicon dioxide, have similar properties.

Structure, bonding, and properties

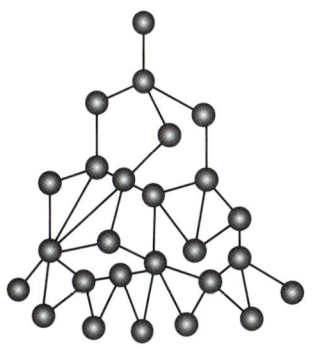

▲ Part of the structure of diamond. The pattern is repeated many times. Each sphere shows a carbon atom and each line shows a covalent bond. In diamond itself, the atoms touch each other.

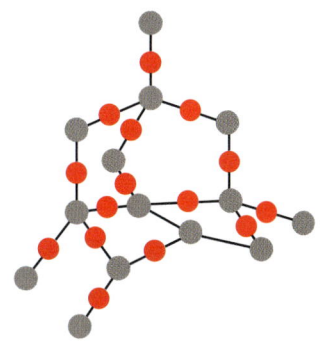

▲ A small part of the structure of silicon dioxide. In silicon dioxide itself, the atoms touch each other.

▲ A crystal of quartz, which is a type of silicon dioxide.

Thinking and working scientifically

Making conclusions from data

Substances with simple molecules have low boiling points. Substances with giant molecular structures have high boiling points.

Grace wants to know the melting points of four covalently bonded substances. She cannot measure the melting points herself as she does not have the substances or suitable apparatus. Instead, she finds the data in a reliable secondary source – a data book.

Substance	Melting point (°C)
carbon monoxide	−205
diamond	3550
silicon dioxide	1710
sulfur dioxide	−72

Grace writes a conclusion from the data:

The substances with low melting points are sulfur dioxide and carbon monoxide. They have simple molecules. The substances with high melting points are diamond and silicon dioxide. They have giant covalent structures.

Questions

1. Write the definition for *giant covalent structure*.
2. a. Name the type of bond that is in substances with simple molecules and giant covalent structures.
   b. Explain why atoms form this type of bond.
3. Explain why a substance with simple molecules has a lower melting point than a substance with a giant covalent structure.

Key points

- A giant covalent structure is a three-dimensional network of atoms that are joined together by covalent bonds.
- Substances with simple molecules have low melting points.
- Substances with giant covalent structures have high melting points.

181

10.7 More about structures

The Burj Khalifa, Dubai (pictured below left), is the tallest building in the world. The tower is made of steel bars, embedded in concrete. Why was steel chosen?

Objectives

- Describe giant metallic structures
- Summarise different types of structure
- Link the structure of an element to its position in the periodic table

▲ Burj Khalifa

Giant metallic structures

As you know, steel is an alloy. It is mainly iron, with small amounts of other elements. Metal elements and alloys are strong because of their structure.

Chapter 3 shows metal atoms arranged in layers. In fact, the atoms have lost their outer electrons to achieve a stable electron configuration. This means that a metal structure is made up of two types of particle:

- positively charged ions
- negatively charged electrons.

The ions are in fixed positions. The electrons move around, between the ions. Electrostatic attraction between the positive ions and negative electrons holds the metal together. This is a **giant metallic structure**.

Key:
− electron
+ metal ion

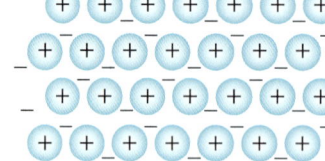

◀ A giant metallic structure has positive metal ions held together by electrons moving around between the ions.

Metallic bonding and physical properties

The strong electrostatic attraction between fixed positive ions and moving negative electrons gives metals these properties:

- high strength
- high melting points.

The moving electrons allow metals to conduct electricity.

182

Structure, bonding, and properties

Summarising structures

The table summarises the four types of structure you have learnt about.

Name of structure	Giant or simple?	Particles in structure	Elements or compounds?	Typical state at 20 °C	Examples
ionic	giant	positive ions and negative ions	compounds	solid	sodium chloride
metallic	giant	positive ions and negative electrons	elements	solid	gold
giant covalent	giant	atoms (joined with covalent bonds)	elements and compounds	solid	carbon (diamond), silicon dioxide
simple covalent	simple	atoms (joined with covalent bonds)	elements and compounds	liquid or gas	hydrogen, carbon dioxide

Structure and the periodic table

The periodic table below shows the type of structure of each element at 20 °C. As you can see, the metal elements have giant metallic structures. Most non-metal elements exist as molecules. A few non-metal elements have giant covalent structures. You can use knowledge about the structure of an element to predict its properties.

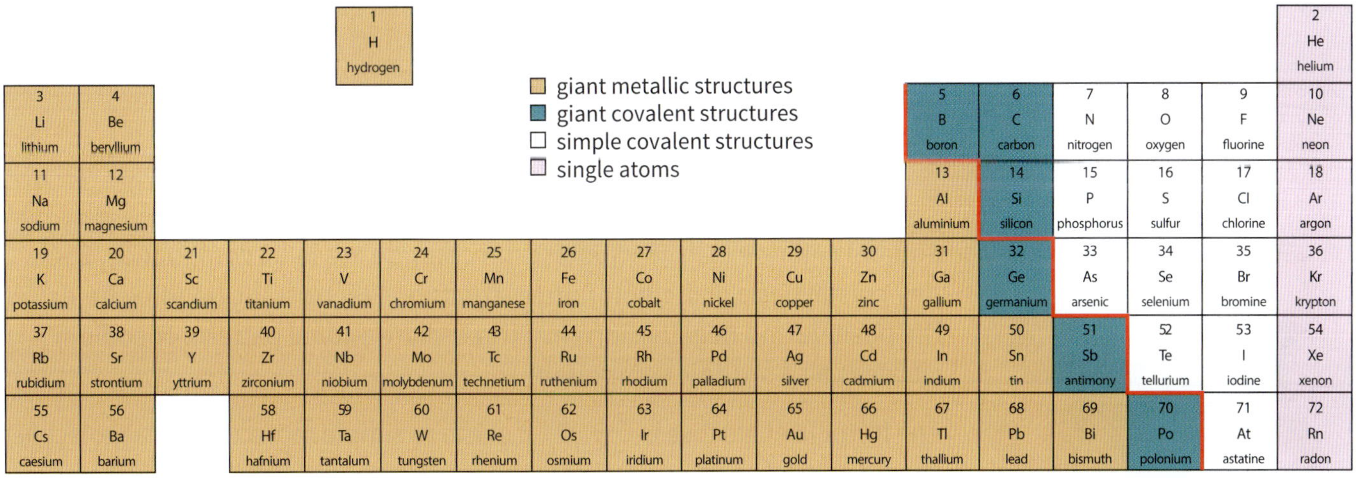

Legend:
- giant metallic structures
- giant covalent structures
- simple covalent structures
- single atoms

Note: this periodic table does not show all the elements

Questions

1. Name the two types of particle in a giant metallic structure, and the force that holds these particles together.
2. Explain why most metals have high melting points.
3. Compare giant ionic and giant metallic structures. In your answer, write down how the structures are similar and different.
4. Suggest why there is no substance with a giant ionic structure shown on the periodic table.

Key points

- A giant metallic structure is the three-dimensional pattern of positive metal ions held together by moving electrons.
- The structure of an element is linked to its position in the periodic table.

Science in context 10.8

Life-saving compounds

Mixed with water, the contents of packets like this save lives. How?

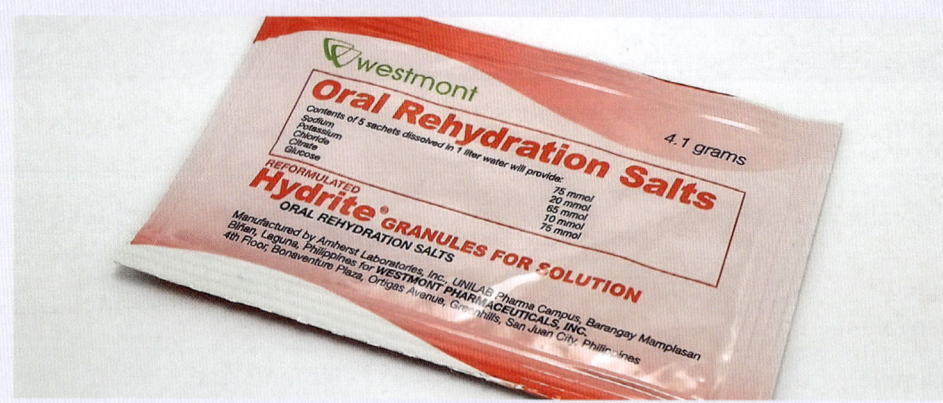

Objectives

- Define the term *systematic review*
- Describe an application of science

Diarrhoea deaths

Diarrhoea takes water out of the body. It removes vital ions, including sodium, potassium, and chloride ions. If the water and ions are not replaced, diarrhoea may cause dehydration. Badly dehydrated people may die.

What's in the sachet?

The sachet contains oral rehydration salts (ORS). There is a mixture of these substances:

- sodium chloride
- potassium chloride
- trisodium citrate
- glucose.

Sodium chloride (salt) and potassium chloride are ionic compounds. Each tiny crystal has a giant ionic lattice structure. Electrostatic bonding holds the positive and negative ions in a three-dimensional pattern. Trisodium citrate is also an ionic compound.

The formula of glucose is $C_6H_{12}O_6$. It exists as simple molecules. In a molecule, shared pairs of electrons make up the covalent bonds that hold the atoms together. The atoms in a glucose molecule are arranged as shown on the left.

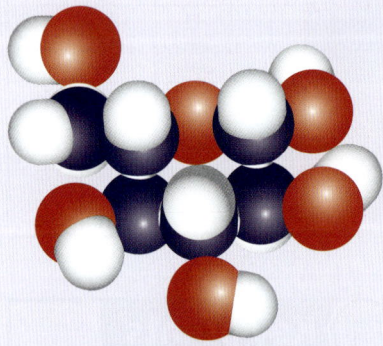

▲ A model of a glucose molecule. It has 6 carbon atoms (shown black), 6 oxygen atoms (red) and 12 hydrogen atoms (white). Three of the hydrogen atoms are difficult to see in this model.

◀ Using oral rehydration solution (ORS). The solution is orange because orange flavouring and colouring have been added.

Saving lives

When you add ORS mixture to water, it dissolves. The solution contains sodium, potassium, chloride, and citrate ions. It also contains dissolved glucose.

Drinking the solution replaces lost water and ions. Glucose helps the body to absorb sodium ions and water. Glucose also provides energy.

Thinking and working scientifically

Investigating oral rehydration solution

In 2010, three scientists at Johns Hopkins University, USA, asked this scientific question:

How many deaths does oral rehydration solution (ORS) prevent in children under 5?

The three scientists decided to use secondary data, from many other scientists, to work out the answer to their question. This is a **systematic review**.

The scientists read 404 scientific papers about the effectiveness of ORS. They did not use all of the papers, for these reasons:

- Some did not investigate ORS in young children.
- Some did not describe their methods in enough detail.
- Some did not report enough data.

The scientists ended up with 157 suitable research papers. They compared the data. They analysed it carefully. They made this conclusion:

For every 100 people who die of diarrhoea and do not take ORS, only 7 would die if they had taken ORS.

▲ *Christa Fischer Walker, one of the scientists who carried out the systematic review.*

The scientists point out that ORS is a simple method for saving lives. However, it is not available to everyone who needs it. The scientists recommend finding better ways of distributing the treatment and getting people to use it.

Questions

1. Write the definition for *systematic review*.
2. Name three ionic substances in ORS.
3. Explain how ORS prevents deaths from diarrhoea.

Key points

- Oral rehydration solution (ORS) saves lives by replacing ions and water lost in diarrhoea.
- A systematic review uses repeatable methods to collect and analyse secondary data from many scientists.

Review 10.9

1. Copy and complete the sentences below. Choose words and phrases from the list.

   negative
   neutron
   positive
   electron configuration
   proton number

   a. The sub-atomic particle with no charge is a _____ [1]

   b. The arrangement of electrons in an atom is its _____ [1]

   c. The number of protons in an atom of an element is its _____ [1]

   d. An ion formed when an atom loses an electron has a _____ charge. [1]

   e. An ion formed when an atom gains an electron has a _____ charge. [1]

2. The table shows the numbers of protons, neutrons, and electrons in the atoms of some elements.

Atom of...	Number of protons	Number of neutrons	Number of electrons
fluorine	9	10	9
sodium	11	12	11
chlorine	17	18	17

   a. Give the proton number of fluorine. [1]

   b. Draw and label a diagram of a fluorine atom nucleus. [2]

   c. Draw the electron configuration of each atom shown in the table. [3]

   d. Use your answer to part **c** to deduce the two elements that are in the same group of the periodic table. Give a reason for your decision. [2]

3. Copy and complete the table below. Use the periodic table to help you. [12]

Element	Chemical symbol	Proton number	Number of electrons in outer shell
boron		5	
	C		
lithium			1
magnesium			2
	Na		

4. Use the periodic table to help you to answer this question.

   a. Give the proton number of aluminium, Al. [1]

   b. Draw the electron configuration of aluminium. [2]

   c. Give the name of one other element that has the same number of electrons in its outer shell as aluminium. [1]

5. Write the chemical formula for each ion below. Use the periodic table to find the chemical symbols of the elements.

   a. A lithium ion, with a charge of +1. [1]

   b. A calcium ion, with a charge of +2. [1]

   c. An iron ion, with a charge of +3. [1]

   d. A chloride ion, with a charge of −1. [1]

   e. A nitride ion, with a charge of −3. [1]

6. The electron configuration of an oxygen atom is shown below. Predict the charge on an oxide ion, and explain your prediction. [2]

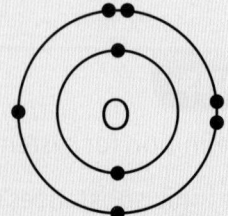

Structure, bonding, and properties

7. The picture shows the ions in magnesium iodide. The purple spheres model iodide ions. The green spheres model magnesium ions.

   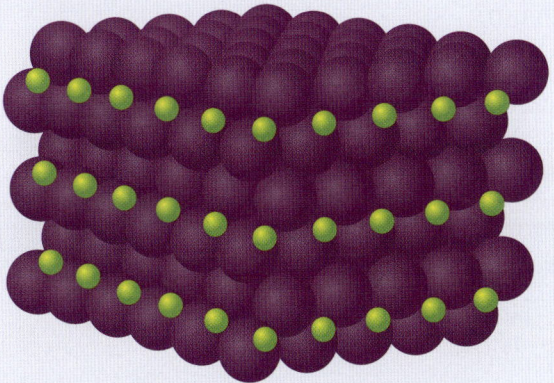

   a. Name the type of attraction between the ions in magnesium iodide. [1]

   b. Explain why magnesium iodide has a high melting point. [1]

   c. A magnesium ion has a charge of +2. Write its formula. [1]

   d. An iodide ion has a charge of −1. Write its formula. [1]

   e. Magnesium iodide has two iodide ions to every one magnesium ion. Write the chemical formula of magnesium iodide. [2]

   f. Explain why magnesium iodide is brittle. [1]

8. The diagram shows the bonding in a fluorine molecule.

   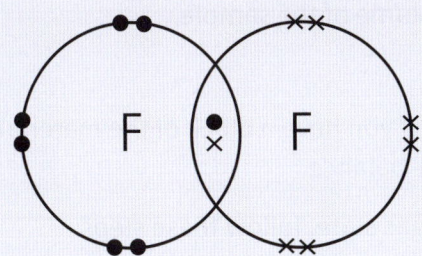

   a. Deduce the formula of a fluorine molecule. [1]

   b. Name the type of bond between the two atoms in a fluorine molecule. [1]

   c. Explain why fluorine has low melting and boiling points. [1]

9. The picture shows a model of a hydrogen sulfide molecule. The yellow sphere represents a sulfur atom.

   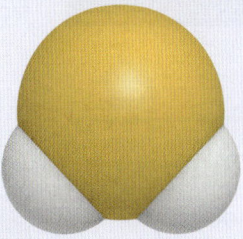

   a. Deduce the formula of hydrogen sulfide. [1]

   b. The atoms in a hydrogen sulfide molecule are joined by covalent bonds. Predict the state of hydrogen sulfide at room temperature. [1]

   c. A sulfur atom has 6 electrons in its outer shell. A hydrogen atom has 1 electron. Draw a dot-and-cross diagram for a hydrogen sulfide molecule. [1]

10. The diagram shows part of a diamond. Each line represents a covalent bond. Each sphere is a carbon atom. In diamond itself, the atoms touch each other.

    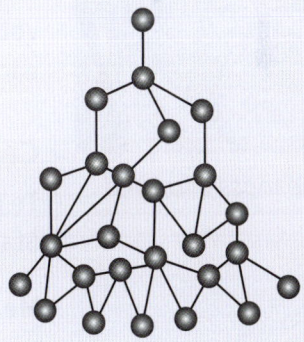

    a. Name the type of structure shown in the diagram. [1]

    b. Explain why diamond has high melting and boiling points. [1]

11.1 Calculating density

Su is a weightlifter. Her weight is made from iron. Why not make weights from another metal, aluminium?

Objectives

- Define the term *density*
- Calculate the density of materials in the solid, liquid, and gas states

▲ Cubes of different metals have different densities.

What is density?

Su's iron weights are heavy. Their mass is 20 kg. Aluminium weights of the same size are less heavy. Their mass is only 7 kg. The iron weights have a greater mass because iron has a greater density than aluminium.

The **density** of a material is its mass in a certain volume. As you know:

- Mass is the amount of matter (stuff) in an object. It is measured in grams (g) or kilograms (kg).
- Volume is the amount of space an object takes up. It is measured in cubic centimetres (cm^3), cubic metres (m^3), or litres (*l*).

Each cube in the picture (above left) is made from a different metal. The volume of each cube is the same. But each metal has a different density, so each cube has a different mass.

Calculating density

To calculate the density of a material, you need a sample of the material. Start by finding the mass and volume of the sample.

▲ Finding the mass of a solid block.

Measuring mass

Use a balance to measure mass. If you have a block of the material, you may be able to place it directly on the balance.

If the material is in the liquid or gas state, follow these steps:

1. Find the mass of the container.
2. Add the material.
3. Find the mass of the container + material
4. Do this calculation:
   mass of material = (mass of container + material) − (mass of container)

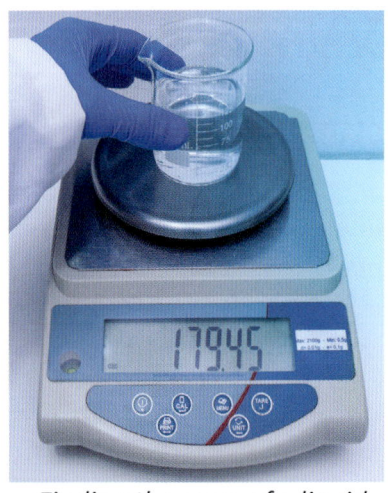

▲ Finding the mass of a liquid.

Measuring volume

If the material is in the liquid state, find its volume with a measuring cylinder. To find the volume accurately:

- Measure the reading at the *bottom* of curved surface of the water.
- Look straight at the scale from the side, not from above or below.

If you have a block of the material, find its volume like this:

Volume = width × length × height
= 6 cm × 10 cm × 9 cm
= 540 cm³

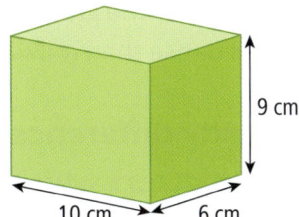

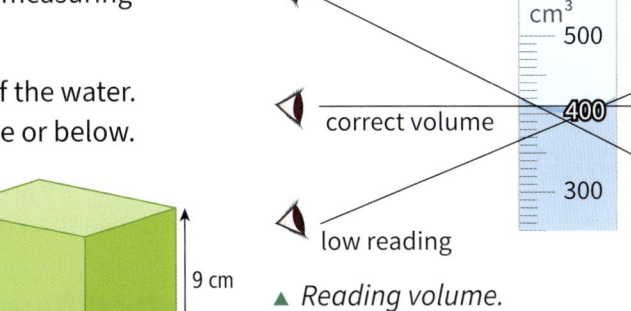

▲ *Reading volume.*

If you have a sample of a solid material that does not have a regular shape, like a stone, find its volume like this:

- Partly fill a measuring cylinder with water to a known volume.
- Add the stone.
- Measure the new volume.
- Calculate the volume increase. This is the volume of the stone.

Calculating density

To calculate density, use this equation:

$$\text{density} = \frac{\text{mass}}{\text{volume}}$$

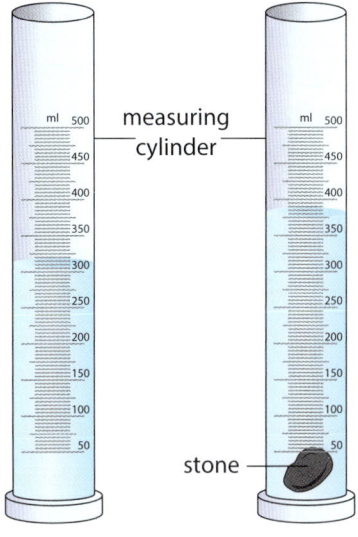

Example 1

A block of wood has a mass of 270 g and a volume of 540 cm³. Calculate its density.

$$\text{density} = \frac{\text{mass}}{\text{volume}} \quad \text{density} = \frac{270\,g}{540\,cm^3}$$

density = 0.5 g/cm³

Example 2

A room has a volume of 30 m³. The mass of air in the room is 36 kg. Calculate the density of the air.

$$\text{density} = \frac{\text{mass}}{\text{volume}} \quad \text{density} = \frac{36\,kg}{30\,m^3}$$

density = 1.2 kg/m³

The next pages explain why different materials have different densities. They also show how ideas about density are useful.

Questions

1. Write the definition for *density*.
2. A silver ring has a mass of 20 g and a volume of 2 cm³. Calculate the density of silver.
3. 20 cm³ of cooking oil has a mass of 18 g. Calculate the density of the cooking oil.
4. A block of iron has sides of these lengths: 2 cm, 3 cm, and 4 cm. The mass of the block is 192 g. Calculate the density of iron.

Key points

- The density of a material is its mass in a certain volume.
- Calculate density using the formula

$$\text{density} = \frac{\text{mass}}{\text{volume}}$$

11.2 Explaining density

Objectives

- Explain why different substances have different densities
- Explain why a substance has different densities in its three states
- Use an analogy to explain density

The glass contains cooking oil, milk, and honey. Why are the layers separate?

The layers are separate because the oil, milk, and honey have different densities. Honey, at the bottom, has the greatest density. Oil, at the top, is less dense than milk and honey.

Density differences

The density of a substance depends on two factors:

- the mass of its particles
- how closely packed its particles are.

Particle mass

In the solid state, substances with the heaviest particles have the greatest densities. The table shows the densities of three metal elements. All the values are given to 2 significant figures.

Metal	Relative mass of atoms	Density (g/cm^3)
magnesium	24	1.8
aluminium	27	2.7
gold	197	19

Particle packing

The particles of a substance in the liquid state are more closely packed than the particles in the gas state. The substance has a greater density in the liquid state than in the gas state.

For most substances, the density of a substance in the solid state is a little greater than its density in the liquid state. This is because its particles are a little more closely packed in the solid state.

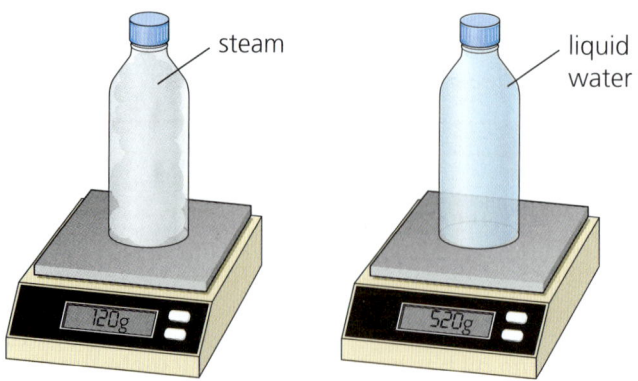

▲ 500 cm^3 of steam has a smaller mass than 500 cm^3 of liquid water. This is because the particles are further apart in steam.

Patterns in the periodic table

Thinking and working scientifically

Modelling density

You can use dried peas to explain why a substance has different densities in its three states. The peas are an analogy for the particle model. As you know, an analogy is a comparison between one thing and another that helps to explain something.

- In the photo, the peas touch each other. They are not in a pattern. If you move the jar gently, the peas move around randomly, sliding over each other. This is an analogy for the liquid state.
- If you shake the jar hard, the peas move far apart. This is an analogy for the gas state.

A volume of 100 cm^3 in the gas model has fewer particles than the same volume in the liquid model. This gives 100 cm^3 of the gas model less mass than 100 cm^3 of the liquid model. The substance has a lower density in the gas state.

The main strength of this analogy is that it helps to explain density. One limitation is that, in the model, you need to move the jar to move the particles. In reality, the particles move themselves.

▲ *Peas in a jar*

Questions

1. Give two factors that affect density.
2. Use the table opposite to state the densities of aluminium and gold, including units. Explain the difference in density.
3. Explain why a substance has a greater density in the liquid state than in the gas state.
4. The mass of a chromium atom is 52. The mass of a tungsten atom is 184. Predict which of the two metals has the higher density. Explain your prediction.
5. Jodie has some chocolates. Suggest how she could use the chocolates as a model to explain why a substance has a greater density in the solid state than in the gas state.

Key points

- The greater the mass of the particles of a substance, the greater its density.
- Density depends on state. The density of most substances is greatest in the solid state, and least in the gas state.
- Models and analogies have strengths and limitations.

Science in context 11.3

Objective

- Describe how scientists build on others' work to develop science knowledge over time

Using density

For thousands of years, people have appreciated jewellery. They made necklaces and rings from gold and silver. They decorated the necklaces with gemstones, such as red rubies and green emeralds.

The work of al-Biruni

More than 1000 years ago, Abu Rayhan al-Biruni was born in what is now Uzbekistan. He became a great scientist and mathematician. He wrote books on many topics, including astronomy, geology, and medicines. Al-Biruni also built on the work of earlier scientists to design new instruments to make accurate measurements.

Today, most scientists specialise in one area of research. They often work in teams. They write about their findings in scientific papers, which are published in journals. All over the world, scientists read the journals. They may do new research of their own, building on what they have read.

▲ Al-Biruni lived more than 1000 years ago.

Finding density

Quartz and diamond look similar. It is not easy see which is which. Quartz is common and cheap, so a cheating trader could make a lot of money by pretending that a lump of quartz was diamond.

A thousand years ago, Al-Biruni asked himself a question:

How can I identify gemstones reliably?

Al-Biruni wondered if he could use density to identify gemstones. He knew how to measure mass. But gemstones do not have regular shapes. How could he measure their volume?

Al-Biruni designed a piece of apparatus to measure volume accurately. It was a flask with a side arm, a bit like the picture above right.

▲ *Quartz (top) and rough diamonds look similar.*

Al-Biruni filled the flask until water overflowed through the side arm, into the beaker. He emptied the beaker. He put it under the side arm again. Then he lowered the gemstone into the flask. Water flowed into the beaker. He measured the volume of the water, which was equal to the volume of the gemstone.

Al-Biruni used this equation to calculate gemstone densities:

$$\text{density} = \frac{\text{mass}}{\text{volume}}$$

The table shows some of al-Biruni's density values.

Material	Density calculated by al-Biruni (g/cm³)	Modern value for density (g/cm³)
quartz	2.58	2.58
ruby	4.01	4.40
pearl	2.70	2.70

▲ *Al-Biruni's apparatus*

Precision

Al-Biruni made measurements with great precision. It was another 700 years before scientists in Europe could measure density with the same level of precision.

Other properties

Al-Biruni did not only measure density. He built on the work of earlier scientists and worked out how to use other properties to classify gemstones, including hardness, colour, crystal shape, and whether the gemstone splits white light into a rainbow.

Gemstones and modern science

Scientists and jewellers still use a method like al-Biruni's to find gemstone density. They also use other techniques to identify gemstones. They measure how well a gemstone conducts electricity, or how much it changes the direction of light.

Questions

1. Name the gemstone in the table above with the highest density.
2. A gemstone has a mass of 2.00 g and a volume of 0.78 cm³.
   a. Calculate the density of the gemstone.
   b. Use your result, and data from the table above, to suggest the identity of the gemstone.
3. Describe one way in which modern scientists work in ways that are similar to al-Biruni.
4. Describe one way in which modern scientists work differently to al-Biruni.

Key points

- Scientists build on others' work to develop science knowledge over time.

11.4 The periodic table: Group 1

An electric car needs a big battery pack. What metal is in the battery?

Objectives

- Give the position of Group 1 in the periodic table
- Describe and explain the Group 1 pattern in melting points
- Present data, describe trends, and make conclusions

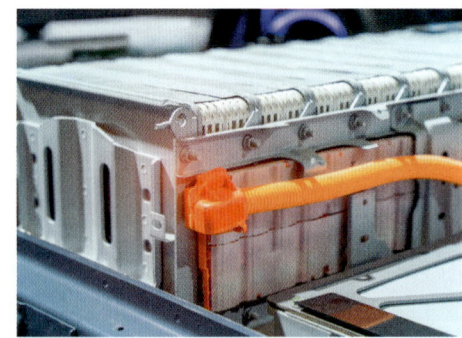

Car batteries rely on lithium. Lithium is a metal element. It is in the column on the left of the periodic table. This column is called **Group 1**. The other Group 1 elements are sodium, potassium, rubidium, caesium, and francium.

																	H													He
Li	Be											B	C	N	O	F	Ne
Na	Mg											Al	Si	P	S	Cl	Ar
K	Ca	Sc	Ti	V	Cr	Mn	Fe	Co	Ni	Cu	Zn	Ga	Ge	As	Se	Br	Kr
Rb	Sr	Y	Zr	Nb	Mo	Tc	Ru	Rh	Pd	Ag	Cd	In	Sn	Sb	Te	I	Xe
Cs	Ba	La	Hf	Ta	W	Re	Os	Ir	Pt	Au	Hg	Tl	Pb	Bi	Po	At	Rn
Fr	Ra	Ac	Rf	Db	Sg	Bh	Hs	Mt	Ds	Rg							

▶ Group 1 is on the left of the periodic table.

▲ As more people buy electric cars, more lithium will be needed. Bolivia has huge reserves of lithium in salt flats like these.

▼ Group 1 metals are soft.

How are Group 1 elements like other metal elements?

The elements of Group 1 are metals. Like other metals, the Group 1 elements:

- conduct electricity
- are shiny when freshly cut.

The Group 1 elements are softer than most other elements. It is easy to cut them with a knife.

Patterns in the periodic table

Thinking and working scientifically

Melting point pattern

Banjeet asks a scientific question:

What is the pattern in melting points for the Group 1 elements?

He cannot measure the melting points himself, so he chooses a trustworthy secondary source. He finds the data. He records the data in a table. The independent variable is in the left column.

Element	Melting point (°C)
lithium	180
sodium	98
potassium	64
rubidium	39
caesium	28

Banjeet plots the data on a bar chart. He chooses a bar chart because the independent variable (the element) is categoric. As you know, a categoric variable is one that is described by words.

Banjeet writes a conclusion:

From lithium, at the top of Group 1, to caesium, at the bottom of Group 1, melting point decreases.

▶ Bar chart showing the melting points of the Group 1 elements.

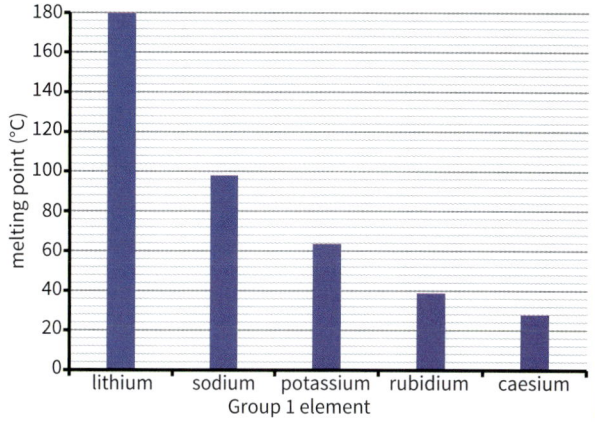

Explaining the melting point pattern

The Group 1 elements have electron configurations like lithium, shown right. From top to bottom of the group, the outer electron gets further from the nucleus.

The Group 1 elements have giant metallic structures. The ions form when each atom loses its one outer electron. The electrons move around, between the ions.

From top to bottom of the group, the positive ions get bigger. This means that electrostatic attractions between positive ions and negative electrons get weaker. As attractions get weaker, ions leave their fixed positions more easily, so melting point decreases.

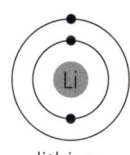

▲ The electron configuration of lithium

▲ The giant metallic structure of a Group 1 element.

Questions

1. Give the names and chemical symbols of the elements in Group 1.
2. Describe the pattern in melting points of the Group 1 elements.
3. Explain the pattern in melting points of the Group 1 elements.
4. The table shows the boiling points of the Group 1 elements.

Element	Boiling point (°C)
lithium	1330
sodium	890
potassium	774
rubidium	688

   a. Plot the data on a bar chart.
   b. Describe the pattern in boiling points.

Key points

- Group 1 is the left column of the periodic table.
- Melting point decreases from top to bottom of Group 1 because electrostatic attractions in the metallic structure decrease.

195

11.5 More about Group 1

Objectives

- Describe the Group 1 pattern in density
- Describe and explain the pattern in the reactions of the Group 1 elements with water

This is an atomic clock. It is accurate to 1 second in 2 million years. The clock relies on the movement of electrons in caesium atoms. In the periodic table, caesium is near the bottom of Group 1.

Patterns in physical properties

As you know, physical properties are properties that you can observe or measure without changing a material. You have seen that the Group 1 elements have patterns in melting point and boiling point. There is also a pattern in another physical property – density.

▲ Bar chart showing the densities of the Group 1 elements.

Density

The table and bar chart show the densities of the Group 1 elements.

Element	Density (g/cm^3)
lithium	0.53
sodium	0.97
potassium	0.86
rubidium	1.53
caesium	1.93

The data show that, overall, density increases from top to bottom of Group 1. The density of potassium does not fit the pattern.

Patterns in chemical properties

Reactions with water

As you know, the chemical properties of a substance describe its chemical reactions. All the Group 1 elements react with water. There is a pattern.

As you saw in Unit 9.7, sodium reacts vigorously with water. The products are sodium hydroxide solution and hydrogen gas.

sodium + water ⟶ sodium hydroxide + hydrogen

The other Group 1 elements also react with water to make a solution of a hydroxide and hydrogen. For example:

lithium + water ⟶ lithium hydroxide + hydrogen

potassium + water ⟶ potassium hydroxide + hydrogen

The reactions get more vigorous from top to bottom of the group.

Explaining the pattern

When a Group 1 element reacts with water, each atom gives its outer electron to an atom of another element. From top to bottom of Group 1, the outer electron gets further from the nucleus. The electrostatic attraction between the nucleus and outer electron gets less, so the outer electron is easier to give away. This makes the reactions more and more vigorous from top to bottom.

Patterns in other groups of the periodic table

Every vertical column in the periodic table is a group. In every group, there is a pattern in physical and chemical properties. The patterns are different for different groups.

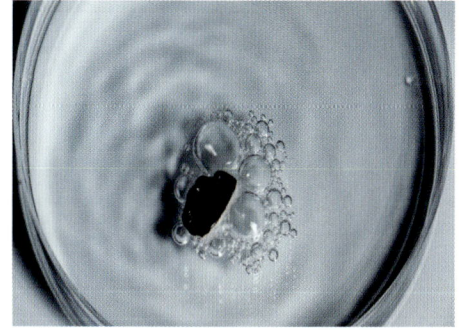
▲ Lithium, at the top of Group 1, reacts quite vigorously with water.

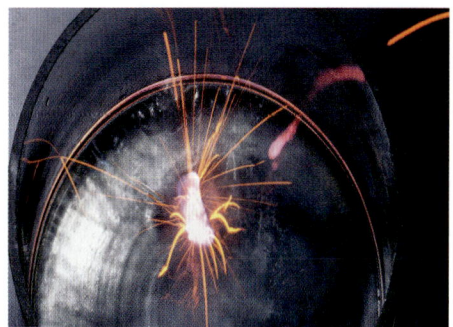
▲ Potassium, lower in Group 1, reacts very vigorously with water.

Questions

1. Describe the pattern in density from top to bottom of Group 1.
2. Predict the products of the reaction of rubidium with water.
3. Explain why the reactions of the Group 1 elements with water get more vigorous from top to bottom of the group.
4. The table shows the relative electrical conductivities of some of the Group 1 elements – the higher the value, the better the element conducts electricity.

Element	Relative electrical conductivity
lithium	11
sodium	21
potassium	14
rubidium	8
caesium	5

   a. Plot the data on a bar chart.
   b. Describe the pattern in electrical conductivity.

Key points

- Overall, density increases from top to bottom of Group 1.
- All the Group 1 elements react with water to make a solution of a hydroxide and hydrogen gas. The reactions get more vigorous from top to bottom of the group.

Extension 11.6

The periodic table: Group 2

Objectives

- Describe patterns in the physical and chemical properties of the Group 2 elements
- Describe an investigation on the reactions of the Group 2 elements

Scientists hope that the James Webb Space Telescope will show us the formation of the first galaxies. Its 18 mirror segments create a mirror of 6.5 m diameter – the biggest ever launched into space.

The mirror segments are made of beryllium, coated with gold. The scientists chose beryllium because it is strong for its weight. It holds its shape well over a range of temperatures.

Group 2 in the periodic table

Beryllium is an element in **Group 2** of the periodic table.

▶ Group 2 is the second column in the periodic table.

Physical properties

Like all groups of the periodic table, the Group 2 elements have patterns in their physical properties. The table (left) shows the melting points of the Group 2 elements.

- the independent variable is the element
- the dependent variable is the melting point (°C)

These data show that, in general, melting point decreases from top to bottom of Group 2. The melting point of magnesium does not fit the pattern.

Group 2 element	melting point (°C)
beryllium	1287
magnesium	650
calcium	842
strontium	777
barium	727
radium	700

▲ Melting points in Group 2.

Chemical properties

As you saw in Unit 10.2, atoms of all Group 2 elements have two electrons in the outer shell. Their similar electron configurations give them similar chemical properties. For example, all Group 2 elements react with dilute hydrochloric acid. The products are a solution of a chloride and hydrogen gas. For example:

strontium + hydrochloric acid ⟶ strontium chloride + hydrogen

Thinking and working scientifically

The Group 2 elements and hydrochloric acid

Catherine asks a scientific question:

What is the pattern in reactivity of the Group 2 elements with hydrochloric acid?

She makes a hypothesis. As you know, a hypothesis is a possible explanation that is based on evidence, and can be tested further.

My hypothesis – From top to bottom of the group, the outer electrons get further from the nucleus, and so easier to give away. This means that the reactions will get more vigorous from top to bottom of the group.

Catherine sets up the apparatus opposite. She makes a prediction:

My prediction – Calcium is below magnesium in Group 2. I predict that the reaction with calcium will be more vigorous.

Catherine collects the gas in the bubbles. She tests the gas with a lighted splint. The splint goes out with a squeaky pop. The gas is hydrogen.

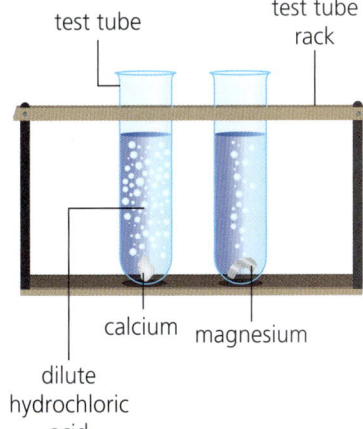

Catherine observes that the calcium reacts more vigorously than magnesium. She makes a conclusion:

My conclusion – The reaction of calcium is more vigorous than the reaction of magnesium. So, for the elements I tested, the reactions get more vigorous going down the group. This is because the outer electrons get further from the nucleus and are easier to give away.

Catherine describes the limitations of her investigation and suggests an improvement.

Limitations – I only tested two Group 2 elements. I could be more certain of my conclusion if I reacted more Group 2 elements with hydrochloric acid.

Key points

- There are patterns in the physical and chemical properties of the Group 2 elements.
- All Group 2 elements react with hydrochloric acid. The products are a solution of a chloride and hydrogen gas. The reactions get more vigorous from top to bottom.

Questions

1. Describe the trend in melting point for the Group 2 elements.
2. Predict the products of the reaction of barium and dilute hydrochloric acid.
3. Write a word equation for the reaction of calcium with dilute hydrochloric acid.
4. Explain why the reactions of the Group 2 elements with hydrochloric acid get more vigorous from top to bottom of the group.

Review 11.7

1. Pawel has a stone. He wants to find its density.

   a. He pours water into a measuring cylinder until it is half full. The diagram shows the surface of the water. Write down the water volume, in cm³. [1]

   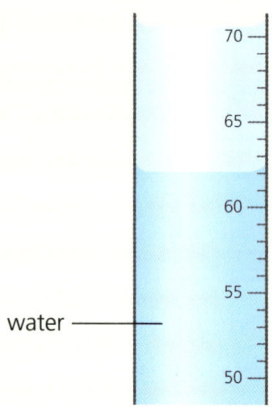

   b. Pawel places the stone in the water. The surface of the water moves up. The diagram shows the new surface of the water.

   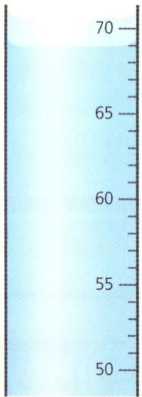

   Use the diagram and your answer to part **a** to work out the volume of the stone. [2]

   c. The mass of the stone is 24 g. Use the equation below to calculate the density of the stone. Include units in your answer.

   $$\text{density} = \frac{\text{mass}}{\text{volume}}$$ [2]

2. The electronic structure of lithium can be written in the form 2,1. This shows that there are two electrons in the first electron shell, nearest the nucleus, and one electron in the outer electron shell, furthest from the nucleus.

   a. Write the electronic structures for sodium and potassium in the same format. [2]

   b. Describe one way in which the electronic structures of lithium, sodium, and potassium are similar to each other. [1]

   c. Describe the link between the atomic structure of an element, and the periodic table group the element is in. [1]

3. The table gives the melting points of four elements in Group 1 of the periodic table.

Element	Melting point (°C)
lithium	180
sodium	98
potassium	64
rubidium	39

   a. Draw a bar chart to show the melting points of the Group 1 elements. [2]

   Label the axes as shown below.

   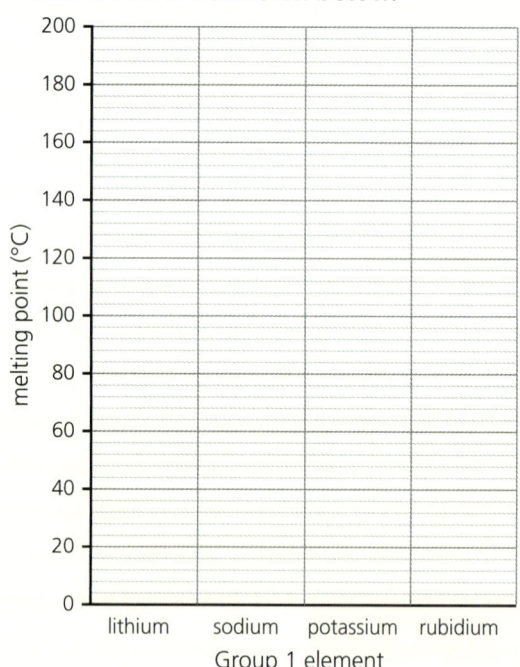

   b. Describe the trend in the melting points of Group 1 elements. [1]

4. Katya watches as her teacher adds a small piece of sodium to water. She makes these observations.

   The sodium moves around on the surface of the water. Bubbles are formed.

After the reaction finished, the teacher added universal indicator to the solution. The indicator went purple.

a. Name the gas in the bubbles. [1]

b. Explain why the indicator went purple. [1]

c. Write a word equation for the reaction of sodium with water. [2]

d. Potassium also reacts with water.

   i. Describe one way in which this reaction is similar to the reaction of sodium with water. [1]

   ii. Describe one way in which the reaction of potassium with water is different from the reaction of sodium with water. [1]

e. Describe the trend (pattern) in the reactions of the first three Group 1 elements (lithium, sodium, and potassium) with water. [1]

5. The table gives the relative sizes of the atoms of the Group 1 elements.

Element	Relative size of atom
lithium	16
sodium	19
potassium	24

a. Describe the trend shown in the table. [1]

b. Use ideas about the electronic structures of the elements to suggest a reason for the trend you described in part **a**. [1]

6. The bar chart shows the melting points of four Group 2 elements. Beryllium is at the top of the group, followed by calcium, strontium and magnesium.

a. Describe the trend shown by the bar chart. [2]

b. Barium is below strontium in the periodic table. Use the bar chart to predict the melting point of barium. [1]

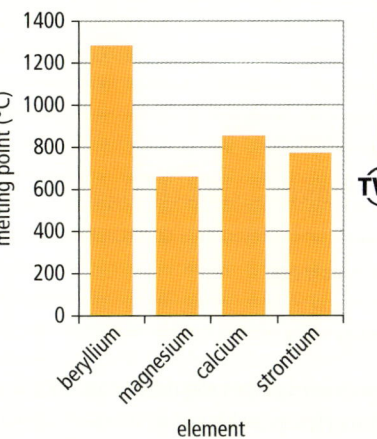

7. Paulo investigates how vigorously the Group 2 elements react with water. He wants to find out if there is a trend in these reactions.

a. Name the independent variable (the variable he changes) and the dependent variable (the variable he observes). [2]

b. Identify one variable Paulo should keep constant. [1]

c. Paulo decides to collect evidence from first-hand experience for the reactions of magnesium and calcium with water. He collects evidence from secondary sources for the reaction of strontium with water. Suggest why he decides not to add strontium to water himself. [1]

8. The Group 7 elements are non-metals.

a. From the list below, choose two properties of the Group 7 elements.

   **do not conduct electricity
   good conductors of heat
   all have high melting points
   all are solid at room temperature
   poor conductors of heat
   good conductors of electricity** [2]

b. The table gives the melting and boiling points of four Group 7 elements.

Element	Melting point (°C)	Boiling point (°C)
fluorine	−220	−118
chlorine		−35
bromine	−7	59
iodine	114	184

   i. Name the element in Group 7 that is liquid at 20 °C. [1]

   ii. Describe the trend in boiling points in Group 7. [1]

   iii. Use the trend in melting points to predict the melting point of chlorine. [1]

201

12.1 Mass and energy in chemical reactions

Objective

- Give definitions for the terms *mass is conserved*, *energy is conserved*, *symbol equation*, and *balancing numbers*

The fuel for the bus is methane. Where does the methane come from?

Bacteria make the methane from human waste. The waste comes from toilets, via a sewage works.

Conserving mass

In the bus engine, there is a chemical reaction. The fuel, methane, reacts with oxygen from the air. The combustion reaction has two products – carbon dioxide and water.

As in all chemical reactions, the atoms rearrange and join together differently. The diagram below models the reaction. Each sphere represents one atom.

Methane…   reacts with…   oxygen…   to make…   carbon dioxide…   and…   water.

   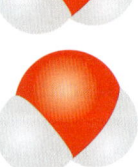

Key:
- carbon atom
- oxygen atom
- hydrogen atom

As you can see, there are the same number of atoms of each element before and after the reaction:

- 1 carbon atom
- 4 hydrogen atoms
- 4 oxygen atoms.

Since the number of atoms does not change, the mass of reactants is equal to the mass of products. The total mass does not change. As in all chemical reactions, **mass is conserved**.

Conserving energy

The combustion of methane is an exothermic chemical reaction. At the start, energy is stored in the methane and oxygen molecules. As methane burns, thermal energy is transferred to the surroundings (in this case, to the bus engine and – in the end – to the air). The total amount of energy does not change. **Energy is conserved**.

Energy is conserved in all chemical reactions, whether they are exothermic or endothermic.

Symbol equations

You can show the reaction of methane with oxygen as a word equation:

$$\text{methane} + \text{oxygen} \rightarrow \text{carbon dioxide} + \text{water}$$

The word equation shows the reactants on the left, and the products on the right. The arrow means 'reacts to make'.

Symbol equations also show chemical reactions. For example:

$$\text{methane} + \text{oxygen} \rightarrow \text{carbon dioxide} + \text{water}$$
$$CH_4 + 2O_2 \rightarrow CO_2 + 2H_2O$$

The symbol equation shows the chemical formula for each substance:

- CH_4 for methane
- O_2 for oxygen
- CO_2 for carbon dioxide
- H_2O for water.

As in the word equation, the reactants are on the left, and the products are on the right. The arrow means 'reacts to make'.

The numbers shown above in red are balancing numbers. **Balancing numbers** show the relative numbers of particles of the reactants and products. The symbol equation above shows that:

- One methane molecule reacts with two oxygen molecules to make one carbon dioxide molecule and two water molecules.

Or that:

- One billion methane molecules react with two billion oxygen molecules to make one billion carbon dioxide molecules and two billion water molecules.

And so on. A balancing number is written to the left of its chemical formula. It is written on the line, and is the same size as the letters in the formula.

> **Key points**
> - Mass is conserved in chemical reactions – the total mass of products equals the total mass of reactants.
> - Energy is conserved in chemical reactions – the total amount of energy does not change.
> - Symbol equations show chemical reactions with chemical formulae.
> - Balancing numbers show the relative numbers of particles of reactants and products.

Questions

1. Write definitions for *mass is conserved* and *energy is conserved* in a chemical reaction.
2. Give the meaning of the arrow in word and symbol equations.
3. Look at the symbol equation for the combustion of methane.
   a. Name the two elements whose atoms are in a methane molecule.
   b. Write the names and formulae of the products of the reaction.
   c. Give the number of oxygen molecules that react with one million methane molecules.

Extension 12.2

Writing symbol equations

Objective

- Write balanced symbol equations for chemical reactions

Darpan mixes two colourless solutions, silver nitrate and potassium iodide. There is a chemical reaction. The reaction makes two products – silver iodide (a yellow precipitate) and potassium nitrate (a colourless solution). Here is the word equation for the reaction:

silver nitrate + potassium iodide ⟶ potassium nitrate + silver iodide

How do you write a symbol equation for the reaction?

Symbol equations

As you know, symbol equations show the reactants and products in a chemical reaction. They also show:

- the chemical formulae of the reactants and products
- the relative numbers of particles of reactants and products
- the states of the reactants and products.

The symbol equation for the precipitation reaction in the photo is:

potassium iodide + silver nitrate ⟶ silver iodide + potassium nitrate

$KI(aq) + AgNO_3(aq) \rightarrow AgI(s) + KNO_3(aq)$

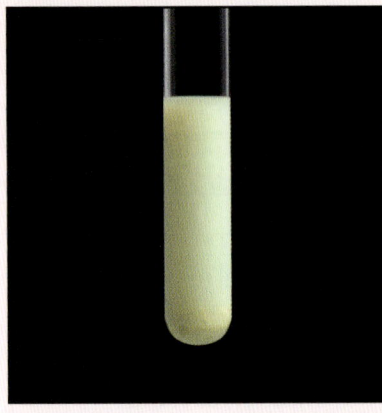

▲ A precipitate of silver iodide

There are no balancing numbers because there are the same numbers of atoms of each element in the reactants and products.

State symbols

The letters in brackets are state symbols. State symbols show the state of the substances in a reaction. They are given below. Write one state symbol to the right of each formula. Do not use capital letters.

- (s) for solid
- (l) for liquid
- (g) for gas
- (aq) for a substance dissolved in water.

Writing symbol equations

Magnesium and oxygen

Magnesium burns brightly in oxygen. The product is magnesium oxide. Follow the steps below to write a balanced symbol equation for the reaction.

▲ Burning magnesium

1. Write a word equation, with the chemical formula under each substance. Do not guess the chemical formulae – look them up, or ask your teacher.

magnesium + oxygen ⟶ magnesium oxide
$Mg + O_2 \rightarrow MgO$

2. Balance the amounts of oxygen. There are two oxygen atoms on the left, and one on the right. Write a big number 2 to the left of MgO. Do not change or add any little numbers.

   $Mg + O_2 \rightarrow \mathbf{2}MgO$

   The big 2 applies to every atom in the formula that follows it. Here it means that there are two magnesium atoms and two oxygen atoms. The equation now shows two oxygen atoms on each side of the arrow.

3. Balance the amounts of magnesium. Write a big 2 on the left of the Mg. The equation now shows two magnesium atoms on each side of the arrow. The equation is balanced.

   $\mathbf{2}Mg + O_2 \rightarrow \mathbf{2}MgO$

4. Add state symbols: $\mathbf{2}Mg(s) + O_2(g) \rightarrow \mathbf{2}MgO(s)$

Zinc and hydrochloric acid

Zinc reacts with dilute hydrochloric acid to make zinc chloride solution and hydrogen gas. Follow these steps to write a balanced symbol equation for the reaction:

1. Write the word equation, with the chemical formula under each substance.

   zinc + hydrochloric acid → zinc chloride + hydrogen
   Zn +   HCl   → $ZnCl_2$ + H_2

2. Balance the amounts of hydrogen. There is one hydrogen atom on the left of the arrow, and there are two hydrogen atoms on the right. Write a big number 2 to the left of HCl.

   $Zn + \mathbf{2}HCl \rightarrow ZnCl_2 + H_2$

3. The equation is now balanced. It shows:
   - one Zn on each side of the arrow
   - two H on each side of the arrow
   - two Cl on each side of the arrow.

4. Add state symbols: $Zn(s) + \mathbf{2}HCl(aq) \rightarrow ZnCl_2(aq) + H_2(g)$

▲ The reaction of zinc with hydrochloric acid.

Questions

Write balanced symbol equations for these chemical reactions:

1. The reaction of solid lithium (Li) with oxygen gas (O_2) to make solid lithium oxide (Li_2O).
2. The reaction of solid magnesium (Mg) with dilute hydrochloric acid (HCl) to make magnesium chloride solution ($MgCl_2$) and hydrogen gas (H_2).
3. The reaction of nitrogen gas (N_2) with oxygen gas (O_2) to make nitrogen monoxide gas (NO).

Key points

To write a balanced symbol equation:

- Write a word equation for the reaction, with the chemical formula below each name.
- Balance the equation by writing balancing numbers where needed.
- Add state symbols.

12.3 Metal displacement reactions

A worker is using a chemical reaction to make liquid iron. The liquid iron cools and freezes, joining the rails together. The reaction is called the thermite reaction.

aluminium + iron oxide ⟶ aluminium oxide + iron

Objectives

- Define the term *metal displacement reaction*
- Predict whether given pairs of substances take part in displacement reactions

Classifying chemical reactions

There are several types of chemical reaction, including combustion, corrosion, and precipitation. The reaction of aluminium with iron is an example of another type of chemical reaction – displacement.

Displacement reactions

The reactivity series

As you know, the reactivity series lists metals in order of how vigorously they react with other substances. Calcium reacts vigorously with oxygen, water, and acids. It is near the top of the reactivity series. Gold, at the bottom of the reactivity series, is inert. It does not take part in chemical reactions.

Displacement reactions involving metal oxides

In a **metal displacement reaction**, a more reactive metal displaces – or pushes out – a less reactive metal from its compound.

The reactivity series shows that aluminium is more reactive than iron. In the thermite reaction, aluminium pushes iron out of iron oxide. The products are aluminium oxide and iron.

In another displacement reaction, magnesium reacts with copper oxide:

magnesium + copper oxide ⟶ magnesium oxide + copper

The particle diagram below models what happens. For each substance, a small part of the giant structure is shown. Each circle represents one atom. The charges on the ions, and the electrons, are not shown.

Part of the reactivity series

calcium
magnesium
aluminium
zinc
iron
copper
silver
gold

▼ *Particle diagram for the reaction of magnesium with copper oxide.*

Key:  magnesium atom
 copper atom
oxygen atom

 +  →  +

magnesium   copper oxide   magnesium oxide   copper

Chemical reactions 3

 Thinking and working scientifically

Carrying out an investigation

Alex heats some pairs of substances. He looks for signs of reaction and observes any products made, and writes his results in a table.

Alex writes word equations for the reactions:

iron + copper oxide ⟶ iron oxide + copper

zinc + copper oxide ⟶ zinc oxide + copper

zinc + iron oxide ⟶ zinc oxide + iron

Metal element	Metal oxide	Observations
iron	copper oxide	glows red, pink-brown metal formed
copper	iron oxide	no reaction
zinc	copper oxide	glows red, pink-brown metal formed
copper	zinc oxide	no reaction
zinc	iron oxide	glows red, silver-coloured metal formed

Alex writes a conclusion:

My results confirm that displacement reactions happen when the metal on its own is higher in the reactivity series than the metal in the compound. If the metal on its own is less reactive than the metal in the compound, there is no reaction.

Displacement reactions in solution

A more reactive metal also displaces a less reactive metal from its compounds in solution.

Mandeep adds magnesium to blue copper sulfate solution. After a few minutes, she sees copper metal in the test tube. The blue solution becomes paler.

There has been a displacement reaction. Magnesium is more reactive than copper, so magnesium displaces copper from copper sulfate solution:

magnesium + copper sulfate ⟶ magnesium sulfate + copper

$Mg(s)$ + $CuSO_4(aq)$ ⟶ $MgSO_4(aq)$ + $Cu(s)$

Later, Mandeep adds copper to magnesium chloride solution. Nothing happens. There is no reaction, because copper is less reactive than magnesium. Copper cannot displace magnesium from magnesium compounds.

▲ *Iron displaces copper from copper sulfate solution.*

Questions

1. Give the meaning of the term *metal displacement reaction*.
2. Decide which of these pairs of substances react. Write word equations for the reactions that occur:
   a. Magnesium and iron oxide
   b. Zinc and magnesium oxide
   c. Zinc and copper sulfate solution
   d. Copper and silver nitrate solution.

Key points

- In a metal displacement reaction, a more reactive metal displaces – pushes out – a less reactive metal from its compound.

Science in context 12.4

Extracting metals

The 55 km Hong Kong-Zhuhai-Macau Bridge is the longest sea-crossing bridge in the world. It is made from steel. Engineers used thousands of tonnes of iron to make the steel. Where did the iron come from?

Objectives

- Define the term *ore*
- Describe the link between the position of a metal in the reactivity series, and how the metal is extracted from its ore
- Describe an application of science

Extracting metals with carbon

In the Earth's crust, most metals are joined to other elements, in compounds. These compounds are mixed with other substances in rocks. A rock that a metal can be extracted from is an **ore**.

Extracting carbon

There are two steps in getting iron from its ore:

1. Separate iron oxide from the compounds it is mixed with.
2. Use a chemical reaction to get iron from iron oxide.

In the chemical reaction, iron is heated with charcoal. Charcoal is a type of carbon.

$$\text{iron oxide} + \text{carbon} \rightarrow \text{iron} + \text{carbon dioxide}$$

▲ This lead statuette was made in Greece over 3000 years ago.

Extracting lead

People have used lead for up to 8000 years. As you know, lead is now used for roofing and to protect from X-rays. Lead exists as lead sulfide in the Earth's crust. It is extracted like this:

1. Heat lead sulfide in air:
   $$\text{lead sulfide} + \text{oxygen} \rightarrow \text{lead oxide} + \text{sulfur dioxide}$$
2. Heat the lead oxide with carbon:
   $$\text{lead oxide} + \text{carbon} \rightarrow \text{lead} + \text{carbon dioxide}$$

Part of the reactivity series

sodium
calcium
magnesium
aluminium
carbon
zinc
lead
iron
copper
silver
gold

Extracting other metals with carbon

Carbon is not a metal. But we can place it in the reactivity series, as shown left. The metals below carbon can be extracted from their oxides by heating with carbon. Carbon is chosen because it is cheap and easy to obtain.

Using electricity to extract metals

The cans below are made from aluminium. Aluminium is above carbon in the reactivity series. In the Earth's crust, it exists as aluminium oxide. Carbon cannot remove oxygen from aluminium oxide. Electricity is used instead. There are two main steps:

1. Dissolve pure aluminium oxide in a special solvent.
2. Pass a 100 000 amp electric current through the solution. The electricity splits up the aluminium oxide. This makes liquid aluminium and oxygen.

◀ Aluminium drink cans

Other reactive metals are also extracted from their compounds by electricity. For example, sodium is extracted by passing an electric current through seawater.

Extracting gold

Gold is at the bottom of the reactivity series. It is unreactive. It is found as an element in the Earth's crust. The metal is easily separated from the substances it is mixed with.

Some gold is found in stream beds, mixed with sand and gravel. You can separate gold by placing the mixture in a pan, and adding water. Gold has a higher density than sand and gravel. It sinks to the bottom of the pan.

◀ Panning for gold

Questions

1. Write the definition for *ore*.
2. Name two metals that are extracted from their compounds by heating with carbon, and two that are extracted using electricity.
3. Give examples to show why extracting metals from their compounds is useful.
4. Predict whether magnesium is extracted from its ore by heating with carbon or with electricity. Give a reason for your prediction.

Key points

- An ore is a rock that a metal can be extracted from.
- Metals above carbon in the reactivity series are extracted from their compounds by electricity.
- Zinc, and the metals below it, are extracted from their compounds by heating with carbon.

Science in context 12.5

Extracting copper

Copper is vital. It makes water pipes, electric wires, and heat exchangers. But how do copper mining and extraction affect the environment?

Objectives

- Describe environmental impacts of copper mining and extraction
- Describe how science applications reduce harmful environmental impacts

Why find new ways of extracting copper?

Environmental impacts

The picture above shows a copper mine. Around the mine are huge piles of waste rock. The land cannot be used for farming or houses. Few plants and animals survive.

Noisy lorries and machinery pollute the air. Getting copper ore out of the ground and extracting copper from the ore need huge amounts of energy. The processes also make large amounts of greenhouse gases, such as carbon dioxide. As you know, extra greenhouse gases in the atmosphere cause climate change.

Increasing demand

The solar cells in this solar farm are connected by copper cables. The wind turbines are connected by undersea copper cables. As more electricity is generated by solar cells and wind turbines, world demand for copper increases.

▲ *Solar cells in the desert*

▲ *Wind turbines in the sea*

Better ways of extracting copper

Scientists are working hard to find better ways to extract copper. Compared with extracting copper from mined copper ore, these methods:

- are less harmful to the environment
- need less energy
- cost less money.

Recycling

Some companies collect used copper objects, such as water pipes and computer parts. They heat the objects to 1085 °C. The copper melts. The liquid copper is poured into moulds. In the moulds, the copper freezes, making ingots. The ingots are sold and made into new objects.

This method of copper recycling has been used for hundreds of years.

Extracting copper from waste

Some devices contain very small amounts of copper. So does copper mine waste. Companies may extract copper from these like this:

1. Spray sulfuric acid onto the waste. This makes copper sulfate solution.
2. Add waste iron to the copper sulfate solution. Iron is higher in the reactivity series than copper, so there is a displacement reaction. The products are iron sulfate solution and solid copper:

iron + copper sulfate $\longrightarrow$ iron sulfate + copper

Phytomining: using plants to extract copper

Some plant species grow well on soil that is mixed with copper ore waste. Copper ions enter plants through their roots. The plants store copper ions in their cells.

Scientists are investigating how to extract copper from these plants. One way is to harvest the crop and burn the plants. The ash is rich in copper compounds.

▲ *Copper waste*

▲ *Copper ingots*

▲ *This plant grows in copper-rich soil.*

Questions

1. Describe three harmful impacts of copper mining and extraction.
2. Explain why the world needs more and more copper.
3. Choose one of these methods: recycling, extracting copper from waste, phytomining. Write a paragraph to explain how the method works and why its environmental impacts are less harmful than extracting copper from mined copper ore.

Key points

- Copper mining has harmful impacts on the environment.
- Copper recycling methods have lower environmental impacts.

12.6 Making salts from acids and metals

The pictures show crystals. What do they have in common?

Objectives
- Define the term *salt*
- Describe how to make a salt from a metal and acid
- Choose suitable equipment
- Do a risk assessment

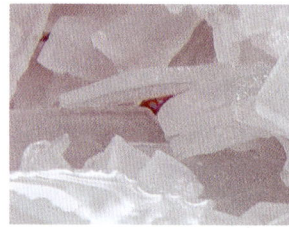
▲ *Magnesium chloride crystals*

▲ *Nickel nitrate crystals*

▲ *Copper sulfate crystals*

The crystals are all salts. A **salt** is a compound made when a metal ion replaces the hydrogen ion in an acid.

Thinking and working scientifically

Making a salt

Seeta is making a salt from magnesium and hydrochloric acid. The equation for the reaction is:

magnesium + hydrochloric acid → magnesium chloride + hydrogen

$$Mg + 2HCl \rightarrow MgCl_2 + H_2$$

Doing a risk assessment

Seeta carries out a risk assessment. She identifies hazards and risks linked to the reactants, the products, and the equipment. She decides how to reduce the chance and consequences of injury from each risk.

Hazard	Risk	Reduce chance of injury and damage by...
magnesium	flammable	keep magnesium away from flames
dilute hydrochloric acid magnesium chloride solution	corrosive – damages eyes and skin	do not touch wear eye protection
hydrogen gas	mixture with air explosive	use small quantities of reactants to make only a small quantity of hydrogen
hot equipment and solutions	burns	wait to cool before touching
breaking apparatus	cuts	wear eye protection

▲ *Flammable hazard symbol*

▲ *Corrosive hazard symbol*

Chemical reactions 3

The chemical reaction

Seeta measures out 25 cm^3 of hydrochloric acid. She uses a measuring cylinder. She does not use a beaker because a measuring cylinder measures smaller differences in volume.

Seeta pours the acid into a beaker. She does not use a conical flask because it is easier to stir the mixture in the beaker.

Next, Seeta adds magnesium to the acid. Bubbles of hydrogen gas form. Soon, the bubbles stop. All the magnesium has reacted. Seeta adds more pieces of magnesium, one by one. She stops when some magnesium remains in the beaker. This shows that all the acid has reacted.

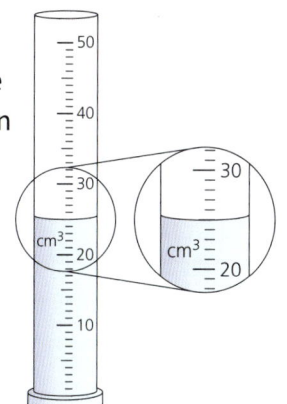

Separating magnesium chloride from the mixture

The beaker contains a mixture. The mixture includes:
- magnesium chloride solution
- solid magnesium.

Filtration

Seeta filters the mixture. She chooses a conical flask, not a beaker, because the conical flask holds the filter funnel upright.

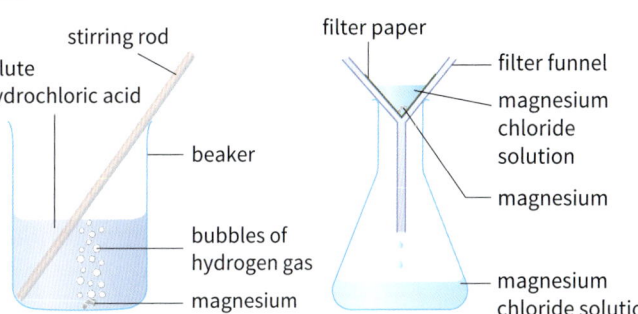

Evaporation

Seeta pours the magnesium chloride solution into an evaporating dish. She chooses an evaporating basin because of its shape – there is a big surface for water to evaporate from.

Seeta heats the solution over a water bath. Water evaporates. After a while, white crystals start to form around the edge. Seeta stops heating.

▲ *Evaporation removes water from the solution.*

Crystallisation

Seeta waits for the evaporating basin to cool.
She moves it to a warm, dry place. The water continues to evaporate, but more slowly. This allows time for crystals to form. Seeta has made her salt, magnesium chloride.

Questions

1. Define the term *salt*.
2. Jay makes a salt from an acid and a metal. Name the process to separate the salt solution from unreacted metal.
3. Write a word equation for the reaction of zinc with hydrochloric acid. Name the salt made.
4. Suggest the metal and acid you could use to make iron sulfate crystals.

Key points

- Make a salt in a chemical reaction. Purify by filtration, evaporation, and crystallisation.
- Do a risk assessment by identifying hazards and risks, and describing how to reduce chance of injury or damage.

213

12.7 More about salts

Objectives

- Choose reactants to make different salts

The picture shows a sculpture. It is made from salts that were dissolved in the waters of Lake Qinghai, in China.

As you know, a salt is a compound made when a metal ion replaces the hydrogen ion in an acid. The salts in the lake exist naturally. You can also make salts in chemical reactions – unit 12.6 describes how to make magnesium chloride, for example.

Making salts – choosing reactants

One of the reactants to make a salt is an acid. Different acids make different salts:

- hydrochloric acid, HCl, makes chlorides
- sulfuric acid, H_2SO_4, makes sulfates
- nitric acid, HNO_3, makes nitrates.

The other reactant must include atoms of the metal element in the salt that is being made. The picture below shows some different salts.

▲ Some salts: sodium manganate, zinc sulfate, copper sulfate, potassium dichromate, nickel chloride, cobalt sulfate.

Chemical reactions 3

Thinking and working scientifically

Making zinc sulfate

Lim wants to make zinc sulfate. He needs two reactants. One of the reactants is sulfuric acid. The other reactant must include zinc. There is a choice of zinc-containing reactants: zinc metal, zinc oxide, or zinc carbonate.

Each of the zinc-containing substances reacts with sulfuric acid to make zinc sulfate. The other products are different:

Reaction 1

zinc + sulfuric acid → zinc sulfate + hydrogen
$Zn + H_2SO_4 \rightarrow ZnSO_4 + H_2$

Reaction 2

zinc oxide + sulfuric acid → zinc sulfate + water
$ZnO + H_2SO_4 \rightarrow ZnSO_4 + H_2O$

Reaction 3

zinc carbonate + sulfuric acid → zinc sulfate + carbon dioxide + water
$ZnCO_3 + H_2SO_4 \rightarrow ZnSO_4 + CO_2 + H_2O$

There are several factors to consider when choosing the metal-containing reactant, including:

- What are the other products of the reaction?
- Is it easy to separate the salt from the other products?
- Is the metal-containing reactant easily available, and what is its cost?

For example, reaction 1 has two disadvantages compared to reactions 2 and 3:

- Zinc is more expensive than zinc oxide and zinc carbonate.
- Hydrogen (made in reaction 1) is flammable, but the products of the other reactions are not.

Lim decides to make his zinc sulfate from zinc carbonate.

Questions

1. Name the acid to make each of these salts in the laboratory:
   a. Nickel chloride
   b. Cobalt sulfate
   c. Sodium nitrate
2. Suggest two reactants that react together to make copper sulfate.
3. A chemist makes nickel chloride from hydrochloric acid and nickel carbonate. Write a word equation for the reaction.

Key points

- A salt is made in a chemical reaction between an acid and a metal-containing substance, for example an oxide or carbonate.

215

Thinking and working scientifically 12.8

Making salts from acids and carbonates

Objectives
- Describe how to make a salt from an acid and an insoluble carbonate
- Evaluate a method and suggest improvements

Fungi can damage crops. Fungicides destroy fungi. Some farmers use fungicides from natural sources, such as the neem tree. Some farmers use fungicides made in factories, such as copper sulfate.

Making copper sulfate

Choosing reactants

Copper sulfate is a salt. It is made in the reactions below:

- **Reaction 1**
  copper + sulfuric acid ⟶ copper sulfate + hydrogen
  The acid must be concentrated and the temperature high.
- **Reaction 2**
  copper oxide + sulfuric acid ⟶ copper sulfate + water
  Dilute acid may be used.
- **Reaction 3**
  copper carbonate + sulfuric acid ⟶ copper sulfate + carbon dioxide + water

Dilute acid may be used.
In industry, copper sulfate is made in reactions 1 and 2. At school, you can use reaction 2 or 3. Reaction 1 is too hazardous.

The chemical reaction

Dan plans to make copper sulfate in this chemical reaction:

copper carbonate + sulfuric acid ⟶ copper sulfate + carbon dioxide + water

$CuCO_3 + H_2SO_4 \rightarrow CuSO_4 + CO_2 + H_2O$

Dan does a risk assessment. Then he starts his investigation.

Dan measures out 25 cm³ of acid. He pours it into a beaker. He adds one spatula measure of copper carbonate powder. Carbon dioxide gas forms, so the mixture bubbles. Dan continues to add copper carbonate, one spatula at a time. He stops when some copper carbonate remains in the beaker, and there is no more bubbling. This shows that all the acid has reacted.

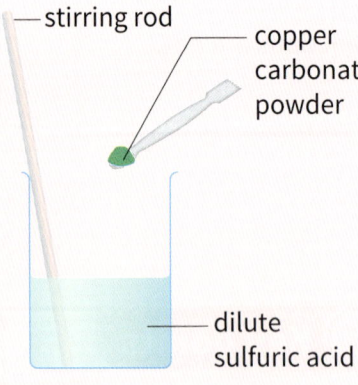

▲ Copper carbonate and sulfuric acid react together to make copper sulfate solution, water, and bubbles of carbon dioxide gas.

Making crystals from the mixture

The beaker now contains a mixture. The mixture includes:

- copper sulfate solution
- copper carbonate powder that has not reacted.

Filtration

Dan filters the mixture. Unreacted copper carbonate remains in the filter paper. He collects copper sulfate solution in an evaporating basin. He chooses an evaporating basin, not a conical flask. This means he does not need to pour the solution from one container to another.

Evaporation

Dan heats the copper sulfate solution. Some of the water evaporates. Some of the solution spits out. After a while, blue crystals start to form around the edge. Dan stops heating.

Crystallisation

Dan waits for the evaporating basin to cool. He moves it to a warm, dry place. The water evaporates slowly. This gives time for crystals to form.

Evaluating and improving methods

Dan makes a smaller mass of copper sulfate crystals than expected. He lost some of the product in the evaporation stage.

Dan wants to improve his method. He repeats the whole experiment. This time, he does not heat the copper sulfate solution over a flame. Instead, he uses a water bath. This heats the solution more evenly. The solution does not spit. Dan makes a greater mass of crystals.

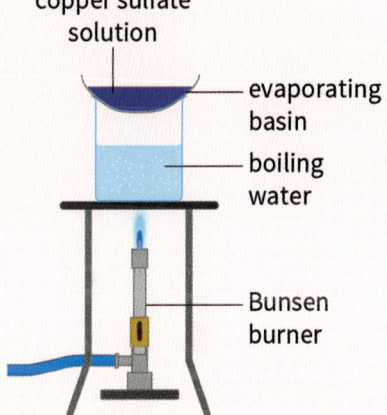

▲ On heating, water evaporates from the copper sulfate solution.

▲ Heating over a water bath.

Questions

1. Write a word equation for the reaction of copper carbonate with dilute sulfuric acid.
2. List the four stages in making a salt from an acid and a carbonate.
3. Explain why carrying out the evaporation step by heating over a water bath is better than heating directly with a Bunsen burner.
4. Write the word equation for the reaction of zinc carbonate with hydrochloric acid. Name the salt made in the reaction.
5. Suggest the metal carbonate and acid you could use to make magnesium nitrate crystals.

Key points

- You can make a salt in a chemical reaction of an acid with an insoluble carbonate. Then purify by filtration, evaporation, and crystallisation.

12.9 Rates of reaction

Some chemical reactions happen very quickly. Others are much slower.

Objectives

- Define the term *rate of reaction*
- Describe how to follow the rate of a reaction that makes a gas
- Explain how to collect reliable results

▲ Chemical reactions in fireworks are fast.

▲ Rusting reactions are slow.

What is rate of reaction?

The **rate of a reaction** is a measure of how quickly a reactant is used up, or how quickly a product forms. Chemists need to control rates of reactions. They may slow down rusting reactions. They may want to speed up reactions that make useful products, such as soap, fertilisers, or medicines.

Before chemists can control reaction rates, they need to find out how fast a reaction is. You cannot tell how quickly a reaction happens by looking at its equation. You need to do an experiment to find out.

Following a reaction

Vijay wants to find out about the rate of the reaction of magnesium with hydrochloric acid. Here is the equation for the reaction:

magnesium + hydrochloric acid → magnesium chloride + hydrogen

$Mg + 2HCl \rightarrow MgCl_2 + H_2$

Obtaining and presenting evidence

Vijay sets up the apparatus shown. He drops a piece of magnesium into the acid. He sees bubbles. The bubbles contain hydrogen gas. As the hydrogen gas forms, it goes into the gas syringe. The plunger moves out.

Vijay measures the total volume of gas made by the end of each minute. He draws a table for his results:

- The independent variable (time) is in the left column.
- The dependent variable (volume of gas) is in the big right column.

Vijay wants to reduce error and obtain reliable results, so he repeats the investigation three times. The table has space for the three results, and for average values.

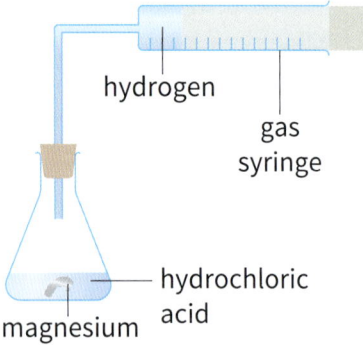

▲ Apparatus to follow the reaction of magnesium with hydrochloric acid.

Time (minutes)	Total volume of gas formed by the end of this minute (cm³)			
	test 1	test 2	test 3	average
0	0	0	0	0
1	31	30	32	31
2	45	47	49	47
3	64	68	66	66
4	69	69	66	
5	76	76	79	77
6	83	85	81	83
7	83	83	83	83
8	84	82	83	83

Vijay chooses how to present his results. The variable he changes, and the variable he measures, are continuous. This means he can plot a line graph:

- The scale for the independent variable is on the *x*-axis.
- The scale for the dependent variable is on the *y*-axis.

Vijay spaces the numbers on the axes evenly.

Vijay draws a cross for each point. He then draws a line of best fit. This is a smooth curve. The number of points above and below the line are equal.

Describing patterns and interpreting results

At first, the graph rises steeply. This shows that hydrogen is formed quickly at the start of the reaction. The rate of the reaction is fast. Then the slope of the graph gets less steep. This shows that the reaction is slowing down. The rate of the reaction is slower.

From the sixth minute onwards, the graph does not go up any more. No more hydrogen gas is being made. This shows that the reaction has finished. All the magnesium has been used up, so there is nothing left for the acid to react with.

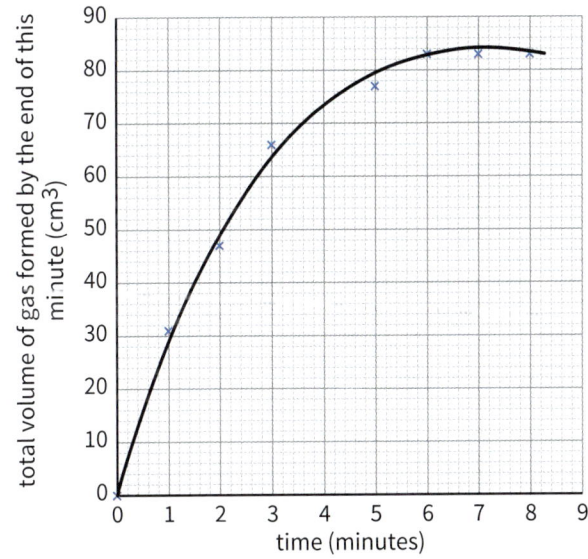

▲ *Graph showing total volume of gas made by the end of each minute.*

Questions

1. Give an example of a fast reaction, and a slow reaction.
2. Use the graph to estimate the volume of gas made during the first 4 minutes of the reaction.
3. Explain how the graph shows when the reaction has finished.
4. Explain why Vijay repeated the experiment three times.
5. Use data in the table to calculate the mean volume of gas collected by the end of the fourth minute.

Key points

In a reaction that makes a gas, the total amount of gas made by the end of each minute shows how the rate of reaction changes with time.

Chemical reactions 3

219

12.10 Concentration and reaction rate

Objective

- Describe and explain how concentration affects reaction rate

Tara is a chemist. She works for a company that makes medicines. The company wants to make its medicines quickly and cheaply. Tara investigates the conditions that speed up reactions that make medicines.

stop watch

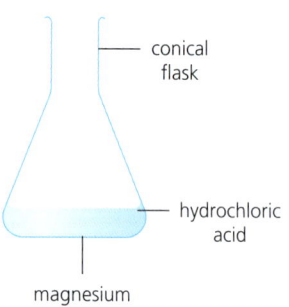

conical flask
hydrochloric acid
magnesium ribbon

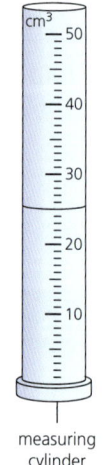

measuring cylinder

▲ Apparatus to investigate the effect of acid concentration on rate of reaction.

Thinking and working scientifically

Planning an investigation

Two students are investigating this scientific question:

> In acid reactions, how does the concentration of acid affect reaction rate?

Their teacher gives them the apparatus shown on the left. The students will use the stopwatch to measure time accurately, and the measuring cylinder to measure volume accurately.

The students list the variables in the investigation:

- concentration of acid
- time for magnesium ribbon to finish reacting
- length of magnesium ribbon
- temperature
- volume of acid

The students think that changing one variable will affect another variable, so they decide to do a fair test. In the test, the students will:

- change the concentration of acid. This is the independent variable.
- measure the time for the magnesium to finish reacting. This is the dependent variable.

The other three variables are control variables. The students will keep these constant, so the test is fair.

The students wonder what range of acid concentration values to use. As you know, range is the difference between the lowest and highest values of a variable. The teacher tells them to use relative concentration values between 1 and 2. The students decide to collect data for five values between these concentrations.

Chemical reactions 3

Presenting evidence

The students carry out the chemical reaction with five concentrations of acid. They write their results in a table. Next, they plot the points on a graph, and draw a line of best fit.

Analysing evidence

The students write a conclusion for their investigation.

The graph shows that, as acid concentration increases, the time for the magnesium ribbon to finish reacting decreases. This means that when the acid concentration is higher, the rate of reaction is faster.

The students need to improve their conclusion by adding a scientific explanation.

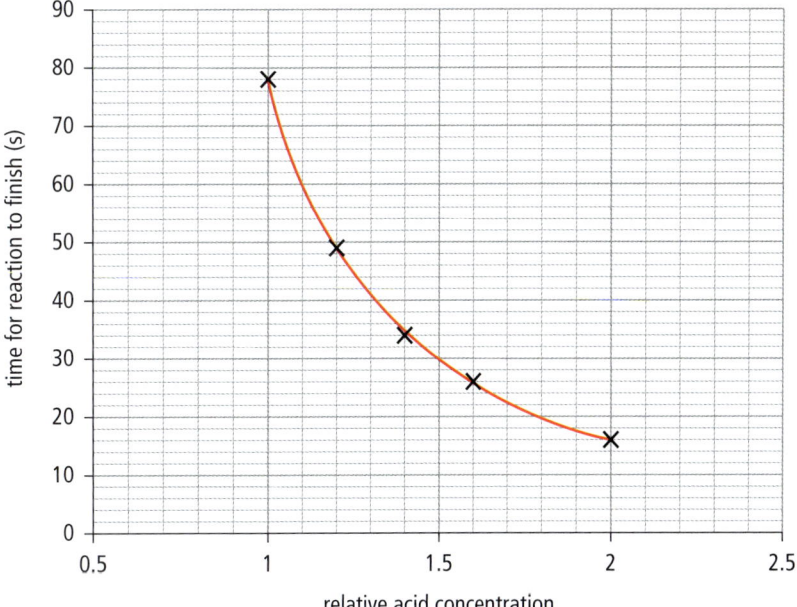

▲ Graph showing the effect of acid concentration on the time for magnesium to finish reacting.

Why does increasing concentration increase reaction rate?

As you know, the concentration of a solution tells you how much solute is dissolved in the solvent. The higher the concentration, the greater the number of acid particles that are dissolved in a certain volume of solution.

Substances can only react when their particles hit each other, or collide. The higher the concentration of solution, the more frequently its particles collide with the other reactant…and the faster the reaction.

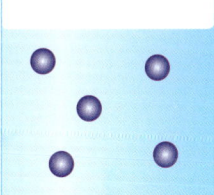
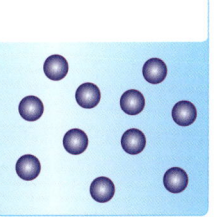

▲ The beaker above shows acid of lower concentration than the beaker below. Water particles are not shown. Not to scale.

Questions

1. Explain why the students displayed their results on a line graph, not a bar chart.
2. a. Describe the relationship between acid concentration and reaction rate.
   b. Use the particle model to explain this relationship.
3. A student wonders whether the results would be similar if they used a different acid. Suggest how they could investigate.

Key points

For reactions involving solutions, the higher the concentration of solution, the faster the reaction.

221

12.11 Temperature and reaction rate

Objective

- Describe and explain how temperature affects reaction rate

Farai and Ibrahim chop potatoes. Ibrahim adds his potatoes to boiling water, at 100 °C. They cook in 15 minutes. Farai adds potatoes to boiling oil, at 200 °C. They cook more quickly.

The chemical reactions that happen when potatoes cook are quicker at higher temperatures. The reaction rates are faster. Is this true for other reactions?

▲ *The higher the temperature, the faster food cooks.*

🌡️ Thinking and working scientifically

Investigating temperature and reaction rate

Farai decides to investigate how temperature affects reaction rate. He thinks that one variable will affect another variable, so he plans a fair test.

Making a hypothesis

Farai makes a hypothesis.

My hypothesis – Reactions happen when particles collide. At higher temperatures, particles move faster and collide more frequently. So, as temperature increases, reaction rate will also increase.

Carrying out an investigation

Sodium thiosulfate solution reacts with hydrochloric acid to make four products:

sodium thiosulfate + hydrochloric acid $\rightarrow$ sodium chloride + water + sulfur dioxide + sulfur

Sulfur is insoluble in water, so it forms as a precipitate. The tiny pieces of solid sulfur make the reaction mixture cloudy.

Farai sets up the apparatus shown. He pours sodium thiosulfate solution into the flask. Then he adds hydrochloric acid. He starts the timer. Gradually, solid sulfur forms. After 244 seconds, the mixture is so cloudy that Farai can no longer see the cross through the flask.

Farai repeats the chemical reaction two more times at the same temperature. He wants to reduce error and obtain reliable results.

Farai makes a prediction:

My prediction – At higher temperatures, the reaction will be quicker, so the cross will disappear after a shorter time.

▲ *The reaction of sodium thiosulfate solution with hydrochloric acid.*

Farai does the same experiment at four different temperatures. Will his prediction be correct?

Presenting results

Farai writes his results in a table.

Temperature of acid (°C)	Time to cover cross (seconds)			
	Test 1	Test 2	Test 3	average (mean)
20	244	240	242	242
30	119	117	118	118
40	59	61	63	61
50	67	32	30	31
60	13	16	16	15

Farai notices that the first result for 50°C is very different to the other results at this temperature. The result is anomalous. He thinks that this is because the acid had cooled down before he used it. He decides not to include this result when calculating the average time for 50°C.

The independent variable (temperature) and dependent variable (time) are both continuous. Farai draws a line graph.

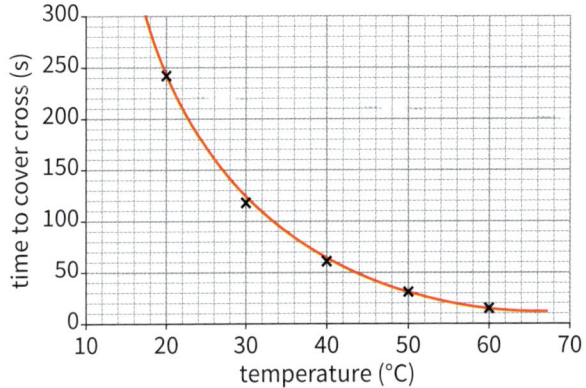

▲ Graph showing time to cover cross at different temperatures.

Analysing evidence

The graph shows that the higher the temperature, the faster the reaction. As temperature increases, so does the rate of reaction.

Farai's prediction is correct. The evidence supports the hypothesis on which he based his prediction.

Why does increasing temperature increase reaction rate?

As you know, substances can only react when their particles collide. At higher temperatures (shown in the box on the far right) particles move faster and collide more frequently. This explains why increasing the temperature increases the reaction rate.

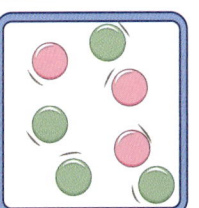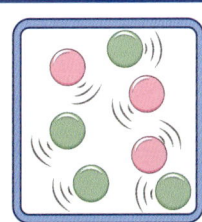

▲ At higher temperatures, particles move faster. There are more frequent and more successful collisions.

Questions

1. Explain why the student repeated his investigation three times.
2. a. Describe the relationship between temperature and rate of reaction.
   b. Use the particle model to explain this relationship.
3. A student wonders whether increasing temperature increases the rate of the reaction of magnesium with hydrochloric acid. Describe an investigation she could do to find out.

Key points

- Increasing temperature increases reaction rate. This is because particles move faster and collide more frequently at higher temperatures.

12.12 Surface area and reaction rate

Objective
- Describe and explain how surface area affects reaction rate

In 1965 in London, England, an explosion at a flour mill killed 4 people and injured 31. Why does flour explode?

Flour has tiny grains. In the air, the grains spread out. Particles from the air – including oxygen molecules – surround the flour grains. Then someone lights a match. A flour grain catches fire. It lights the grains near it. A flame flashes through the flour cloud. This is the explosion. Other powders form explosive mixtures with air, including sugar, sawdust, and aluminium.

▲ A test explosion of aluminium powder.

Thinking and working scientifically

Investigating surface area and reaction rate

Planning an investigation

Saffron wants to investigate reactions involving powders. She asks a scientific question:

My question – Do powders react quicker than big lumps of solid?

Saffron uses her scientific knowledge to make a hypothesis:

My hypothesis – Reactions happen when particles collide. Only the surface particles of a solid can react. A certain mass of powder has more surface particles than the same mass of big lumps, so a powder reacts more quickly than big lumps.

Saffron decides to investigate the reaction of calcium carbonate with hydrochloric acid:

calcium carbonate + hydrochloric acid → calcium chloride + water + carbon dioxide

$$CaCO_3 + 2HCl \rightarrow CaCl_2 + H_2O + CO_2$$

Saffron sets up the apparatus shown.

As carbon dioxide is made, it escapes from the apparatus. The mass of the flask and its contents decreases.

Saffron makes a prediction:

My prediction – The mass will decrease most quickly with the powder.

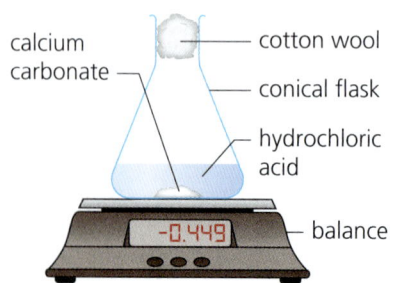

▲ Apparatus to investigate how surface area affects reaction rate.

224

Chemical reactions 3

Carrying out an investigation

Saffron adds a big lump of calcium carbonate to dilute hydrochloric acid. She measures the time for the mass to decrease by 1.0 g.

Saffron does the same experiment two more times – with small lumps of calcium carbonate, and with calcium carbonate powder.

Presenting results

Saffron records her results in a table.

Size of calcium carbonate pieces	Time for total mass to decrease by 1 g (min)
Big lump	10
Powder	1
Small lumps	5

Saffron draws a bar chart because the independent variable (the size of the calcium carbonate pieces) is discrete.

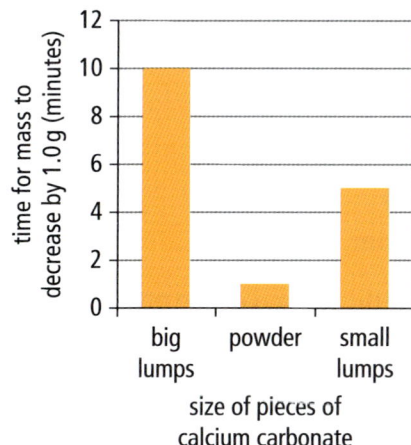

▲ Bar chart showing how surface area affects reaction rate.

Analysing results

Saffron's prediction is correct. The evidence supports the hypothesis on which she based her prediction.

Why does increasing surface area increase reaction rate?

Ten grams of powder has a greater surface area than one 10 g lump of the same substance. There are more particles on the surface of the powder than on the surface of the lump.

Substances can only react when their particles collide. If a substance is in the solid state, only its surface particles can react. The greater the surface area, the faster the reaction.

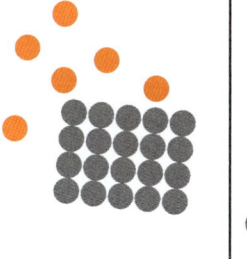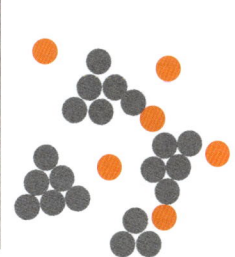

▲ The greater the number of particles on the surface, the faster the reaction.

▲ The sugar in both piles has the same mass. The sugar on the right has the greater surface area.

Questions

TWS 1. Write down three control variables in Saffron's investigation.

2. a. Describe the relationship between surface area and rate of reaction.

   b. Use the particle model to explain this relationship.

3. Write down two other factors that affect the rate of a reaction.

📖 Key points

- Increasing surface area increases reaction rate. This is because the greater the surface area, the greater the number of particles available for reaction.

Review 12.13

1. This question is about displacement reactions. The list below shows part of the reactivity series. Use it to help you answer the question.

   zinc
   iron
   lead
   copper

   a. Predict which of the pairs of substances below will react. [2]

      i. Copper and zinc oxide

      ii. Lead and copper oxide

      iii. Zinc and lead oxide

      iv. Lead and iron oxide

   b. Write word equations for the pairs of substances in part **a** that react. [4]

2. Caz puts a piece of zinc in some copper sulfate solution. A reaction takes place.
   The word equation for the reaction is:
   zinc + copper sulfate → copper + zinc sulfate

   a. Name the products of the reaction. [1]

   b. Explain why the reaction is a displacement reaction. [1]

   c. Caz places a piece of zinc in some nickel nitrate solution. A displacement reaction takes place.

      i. Explain whether nickel or zinc is the more reactive metal. [2]

      ii. Write a word equation for the reaction. [2]

      iii. Predict what would happen if Caz placed a piece of nickel in zinc chloride solution. Give a reason for your prediction. [2]

3. Name the salts made when the following pairs of substances react.

   a. Magnesium and sulfuric acid [1]

   b. Zinc and hydrochloric acid [1]

   c. Magnesium and nitric acid [1]

   d. Copper carbonate and hydrochloric acid [1]

   e. Zinc and sulfuric acid [1]

4. Lia plans to make zinc chloride crystals. She decides to react zinc metal with hydrochloric acid.

   a. The products of the reaction are zinc chloride solution and hydrogen gas. Write a word equation for the reaction. [2]

   b. Lia uses a secondary source to list the hazards of the reactants and products.

Substance	Hazard
zinc metal	Low hazard
dilute hydrochloric acid	Low hazard. May cause harm in eyes or in a cut.
hydrogen gas	Extremely flammable. Forms explosive mixture with air.
zinc chloride solution (dilute)	Low hazard
zinc chloride crystals and concentrated solutions	Corrosive. Burn skin. Harmful if swallowed.

   Lia takes the precautions below to reduce risks from the hazards. Give one reason for each precaution.

   i. Be careful not to spill the acid. [1]

   ii. Wear eye protection at all stages of the experiment. [1]

   iii. Do not touch the zinc chloride crystals. [1]

226

c. Lia makes zinc chloride solution. She pours the solution into an evaporating dish. She does not heat the solution directly. Instead, she heats the solution over a water bath. Suggest why. [1]

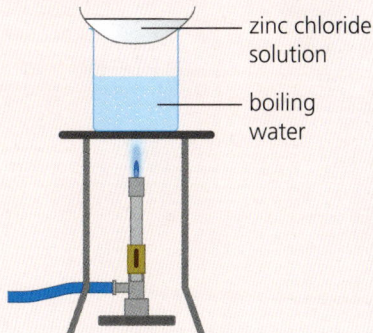

5. Rashid investigates the reaction of hydrochloric acid with calcium carbonate.

The equation for the reaction is:

calcium carbonate + hydrochloric acid →
     calcium chloride + carbon dioxide + water

a. Predict which product of the reaction is formed as a gas. [1]

b. Rashid sets up the apparatus below.

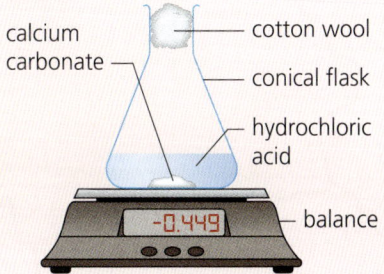

i. Rashid predicts that, as the reaction progresses, the mass of the reaction mixture in the flask will decrease. Which of the reasons below best explains why the mass decreases? [1]

   A The product that is formed as a gas dissolves in the reaction mixture.

   B The product that is formed as a gas escapes into the air.

   C The product that is formed as a gas is made up of atoms of two elements.

ii. Rashid records the mass of the reactant mixture every minute. He plots a graph of his results.
    Predict which of the graphs below best represents Rashid's results. [1]

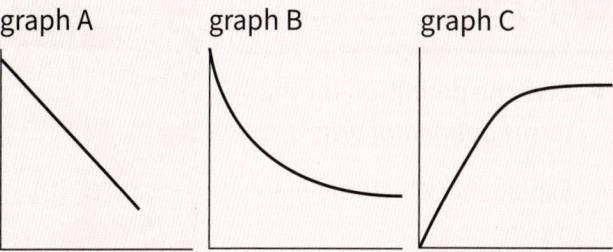

c. Rashid wants to investigate how the size of the pieces of calcium carbonate affect the reaction rate.
   He has three sizes of calcium carbonate pieces – big lumps, small lumps, and powder. He decides to measure the time taken for 1.0 g of gas to be made with each size of calcium carbonate pieces.

i. Name the variable he changes. [1]

ii. Name the variable he measures. [1]

iii. Name two variables Rashid must keep constant. [1]

iv. Rashid finds that the powder reacts most quickly.
    Which of the reasons below best explains why? [1]

   A For a certain mass of calcium carbonate, the powder has the smallest surface area.

   B For a certain mass of calcium carbonate, the powder has the biggest surface area.

   C For a certain mass of calcium carbonate, the powder has the highest concentration.

13.1 Continental drift

Objectives

- Give the definition for the term *continental drift*
- Explain how tectonic plates move

Every year, the gap between the rocks gets wider. Can you explain why?

Tectonic plates

The gap is in Iceland. It separates two tectonic plates. On one side is the North American plate. On the other side is the Eurasian plate.

As you know, the surface of the Earth consists of about 12 tectonic plates. Each is a slab of solid rock. It is made up of a piece of the Earth's crust and the top part of the mantle below it.

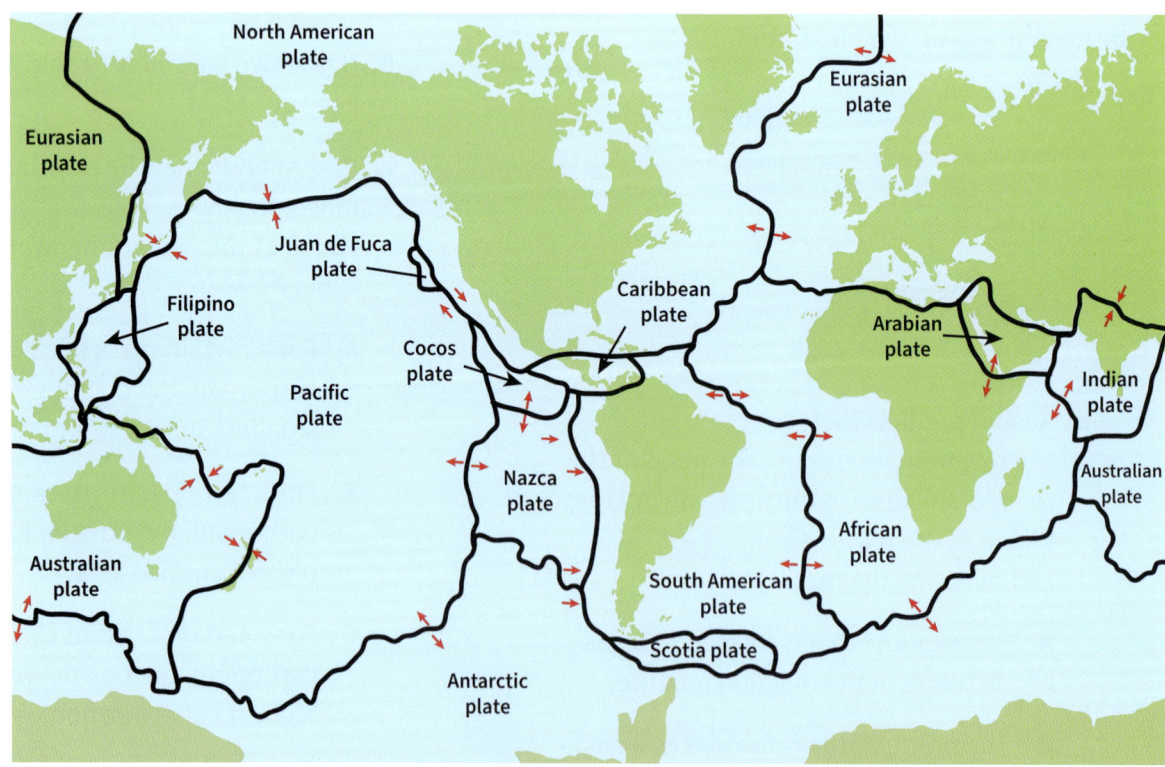

The map shows the Earth's tectonic plates. The arrows show how the plates are moving. The red dots show the positions of earthquakes and volcanoes. They are mostly at plate boundaries, where tectonic plates meet. This is evidence for tectonic plates.

Planet Earth

Continental drift

Why do tectonic plates move?

Tectonic plate rock is less dense than the mantle below it, so the plates rest on the mantle. Deep inside the Earth, natural processes heat the mantle. The heat drives convection currents in the mantle. The convection currents make the tectonic plates move.

Where have tectonic plates moved?

Tectonic plates move slowly, at speeds of a few centimetres a year. Over millions of years, tectonic plates have moved many kilometres. This is **continental drift**.

The maps below show how the continents have moved over the past 225 million years.

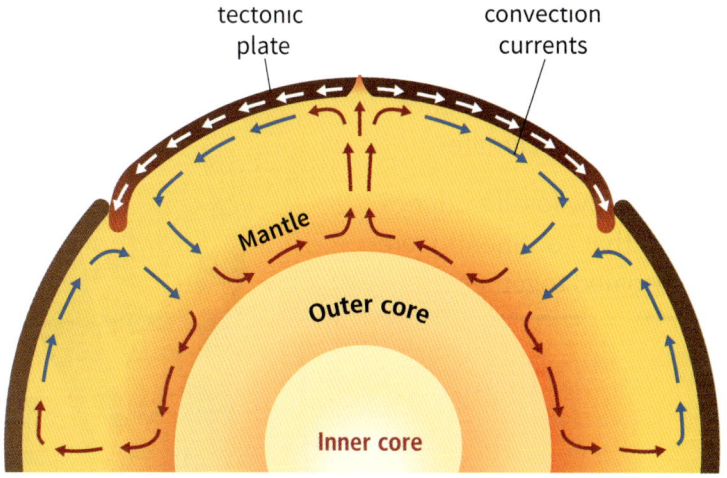
▲ Convection currents in the mantle.

▲ Permian period, 225 million years ago

▲ Jurassic period, 150 million years ago

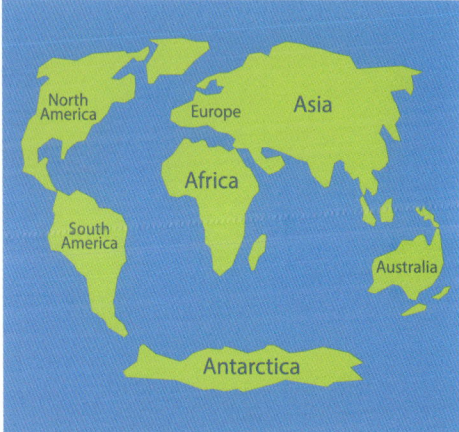
▲ Present day

Questions

1. Give the definition for *continental drift*.
2. Explain what makes tectonic plates move.
3. Look at the map opposite
   a. Name two plates that are moving towards each other.
   b. Name two plates that are moving away from each other.

Key points

- Continental drift is the movement of tectonic plates over millions of years.
- Tectonic plates move as a result of convection currents in the mantle below.

229

13.2 Evidence from fossils

Objectives

- Explain how fossilised trees show that Antarctica once had a warmer climate
- Explain how fossil evidence shows that continents were once joined together

In 1912, Robert Falcon Scott and his team were returning from the South pole. They rested in a rocky area, and searched for fossils. The explorers were excited by their fossil finds. They took the fossils away with them.

Nine months later, a rescue team found the heavy fossils again, next to the explorers' frozen bodies. Why did the explorers keep the fossils? What made them so special?

Evidence from trees

Scott wrote that one of the fossils was:

A piece of coal with beautifully traced leaves in layers, and some excellently preserved impressions of thick stems.

We now know that the fossils are remains of an extinct tree, called *Glossopteris*.

▲ A fossil from the extinct tree species Glossopteris, found in India.

▲ A *Glossopteris* tree probably looked like this. Some were 30 m tall.

Planet Earth

Today, trees do not grow on Antarctica. It is too cold in the area around the South pole. The fossilised leaves are evidence that the land that is now Antarctica once had a warmer climate. The continent must have been further north, away from the South pole.

Before these fossils were found in the Antarctic, people had found fossils of the same tree species, *Glossopteris*, on four continents – Africa, Australia, India, and South America. This suggests that the continents were once joined together.

Evidence from animals

As you know, Alfred Wegener used fossil evidence to support his 1912 hypothesis that the continents were once a single piece of land. The land broke up, he said, and the pieces drifted apart.

Wegener used fossil evidence from several species, including *Glossopteris* and an extinct lizard, *Mesosaurus* (see Unit 6.2). Fossil remains of *Lystrosaurus* also provided evidence for Wegener's hypothesis. The animal had just two teeth, and used its beak to tear leaves from plants.

▲ *A reconstruction of Lystrosaurus. Its fossilised remains have been found in Africa, India, Antarctica, China, and Mongolia.*

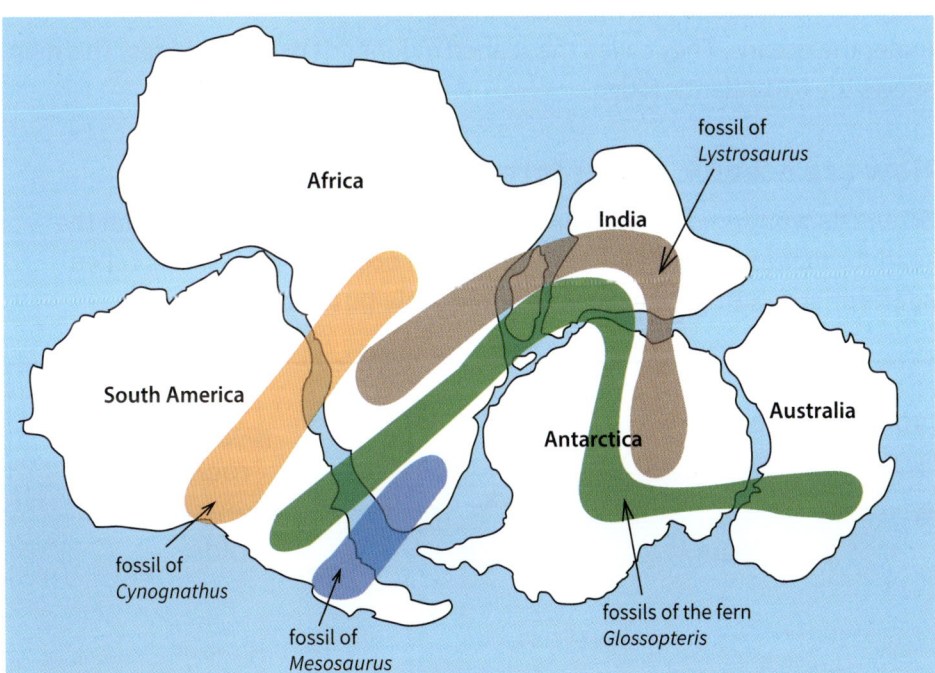

◄ *Each colour on this map shows the areas that fossils of one species have been found in.*

Questions

1. Explain how fossilised trees show that the land that is now Antarctica once had a warmer climate.
2. Look at the map, and name the two continents on which fossil remains of *Cynognathus* have been found.
3. Look at the map. Give evidence from two species that show that India and Africa were once joined together.

Key points

- Fossil evidence suggests that Antarctica was once further from the South pole than it is now.
- Fossil evidence suggests that the continents were once joined together.

13.3 Evidence from seafloor spreading

Objectives

- Define the terms *oceanic ridge* and *seafloor spreading*
- Explain how symmetrical magnetic stripes on the two sides of an oceanic ridge provide evidence for seafloor spreading

The picture shows a seafloor mapping instrument. It is towed by a ship. It sends sound waves to the seafloor. The sound waves are reflected, detected, and processed to make an image of the seafloor. How do seafloor images provide evidence for plate tectonics?

Oceanic ridges

In spite of evidence from fossils and continent shapes, most scientists rejected Wegener's hypothesis at first. They did not know *how* continents could move apart.

The answer came when scientists started to investigate the seafloor in the 1950s. They were amazed to find mountain chains under the oceans. They called the seabed mountains **oceanic ridges**. This map shows the Mid-Atlantic Ridge.

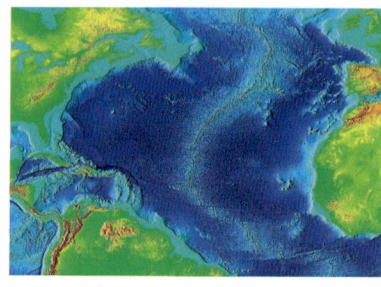

▲ *A mid-oceanic ridge.*

How are oceanic ridges formed?

Scientists wondered how oceanic ridges are formed. It turns out that the seafloor moves away on the two sides of an oceanic ridge. This is called **seafloor spreading**. The diagram shows how this happens.

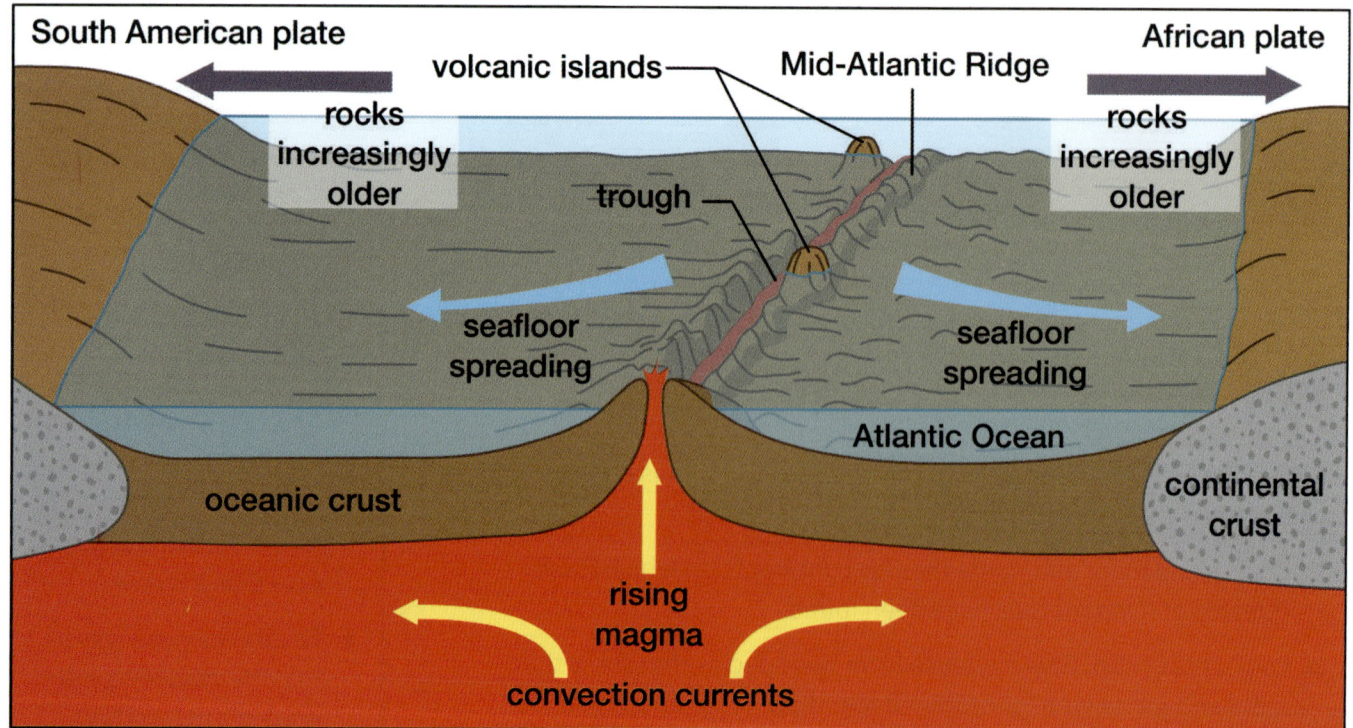

Magnetic stripes

At an oceanic ridge, hot liquid rock – magma – cools and freezes. This makes new rock. Much magma is rich in iron compounds. This magma forms rock that is magnetised in the direction of the Earth's magnetic field.

Every now and again, the Earth's magnetic poles flip. The magnetic North pole becomes the magnetic South pole, and the magnetic South pole becomes the magnetic North pole. Rock formed in these times is magnetised in the opposite direction.

The pattern of symmetrical stripes in the rock on both sides of mid-ocean ridges is evidence for seafloor spreading. By 1966, many scientists knew about this evidence. They understood *how* the continents could drift apart. They accepted the idea of plate tectonics.

▼ *Rock magnetism on the two sides of the oceanic ridge has the same stripe pattern.*

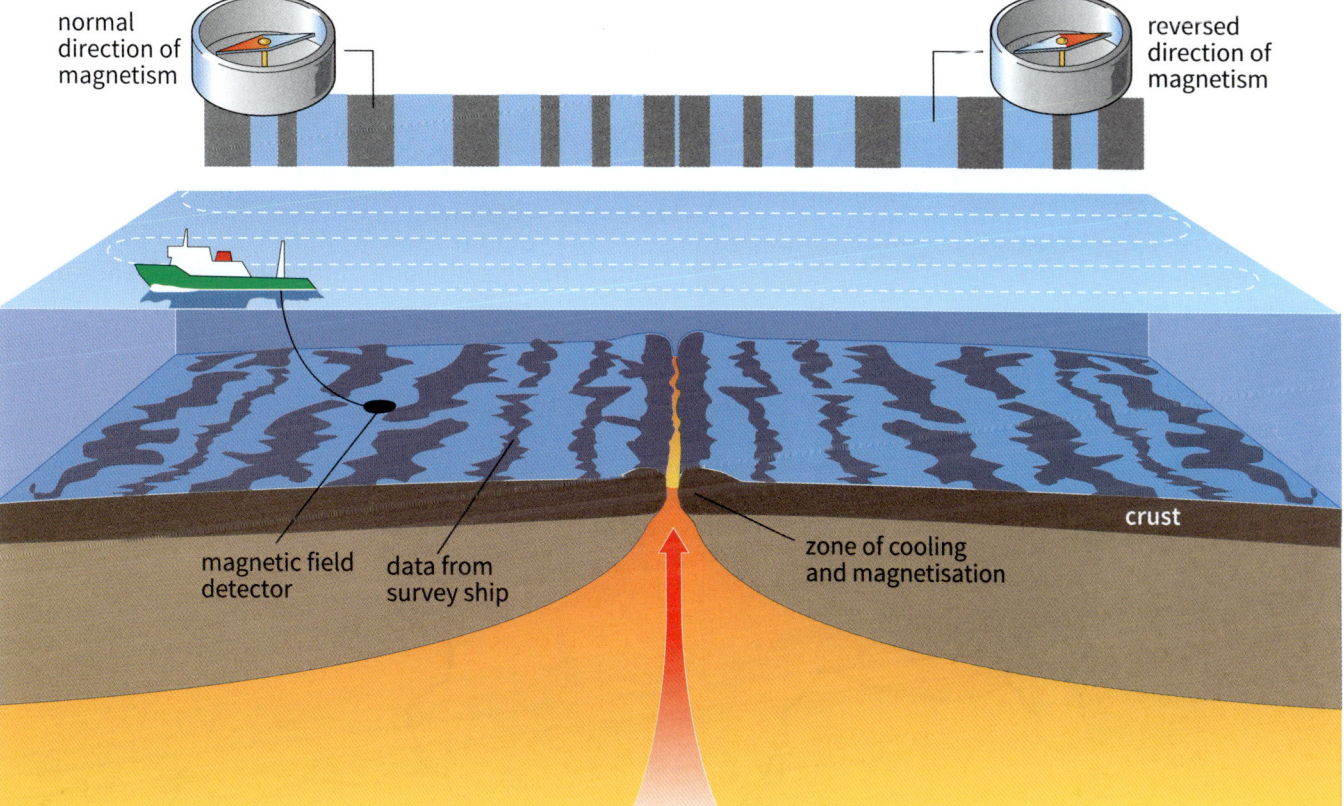

Questions

1. Write definitions for the terms *oceanic ridge* and *seafloor spreading*.
2. Explain how seafloor spreading occurs.
3. Explain why there is a pattern of magnetic stripes on the two sides of a mid-ocean ridge.

Key points

- An oceanic ridge is a mountain chain on the ocean floor.
- Seafloor spreading is the movement of the seafloor away from the two sides of an oceanic ridge.
- The symmetrical pattern of stripes in the rock on the two sides of a mid-ocean ridge is evidence for seafloor spreading.

Review 13.4

You may need to refer to Chapter 6 to help you to answer some of the questions.

1. This question is about tectonic plates.

   a. Give the approximate number of tectonic plates on the Earth. [1]

   b. Give the state of the rock in a tectonic plate. [1]

   c. Compare the density of a tectonic plate to the density of the mantle below it. [1]

   d. Explain how tectonic plates move. [2]

2. This question is about continental drift.

   a. Write the definition for continental drift. [1]

   b. Fossilised leaves have been found in Antarctica.

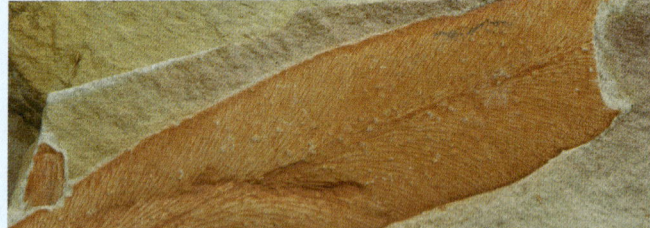

   Explain how the fossils are evidence that Antarctica was once not at the South Pole. [1]

   c. The picture on the stamp shows what an extinct animal, *Cynognathus,* might have looked like.

   Fossilised remains of *Cynognathus* have been found on two continents, South America and Africa.

   i. Explain how the fossils are evidence that South America and Africa were once joined together. [1]

   ii. Describe one other piece of evidence that South America and Africa were once joined together. [1]

3. Use the tectonic plate map below to help you to answer this question.

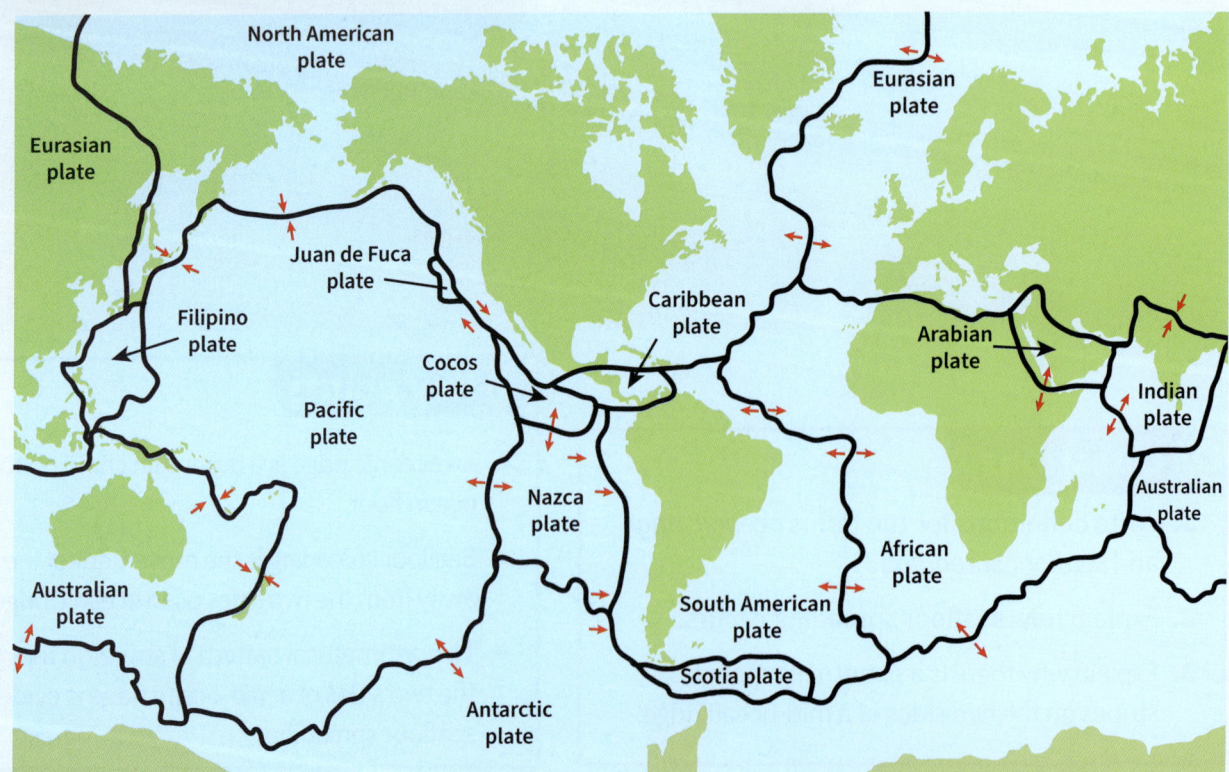

a. Find the Pacific plate and the Nazca plate on the map. State whether the two plates are moving towards each other or away from each other. [1]

b. There is an oceanic ridge on the Pacific–Nazca plate boundary.

  i. Give the definition for an oceanic ridge. [1]

  ii. An oceanic ridge forms from magma that rises up, between the two plates. Name the change of state that happens when the magma cools and becomes solid. [1]

c. Find the South American plate on the map.

  i. State whether the South American plate is moving towards, or away from, the Nazca plate. [1]

  ii. Describe how mountains formed on the West (left) of South America. You may need to look at Unit 6.3 to help you to answer this question. [2]

4. Use the map below to help you to answer this question. The map shows some minor tectonic plates, which are not labelled on the world map above.

a. Explain why the Red Sea is getting wider. [3]

b. There is a mountain chain on the boundary of the Arabian and Eurasian plates. Use information from the map to explain why. [3]

c. Explain why there are frequent earthquakes on the boundary of the Arabian and Turkish plates. [1]

d. Calculate the distance travelled by the Arabian plate in 100 years, assuming its speed does not change. Give your answer in millimetres. [2]

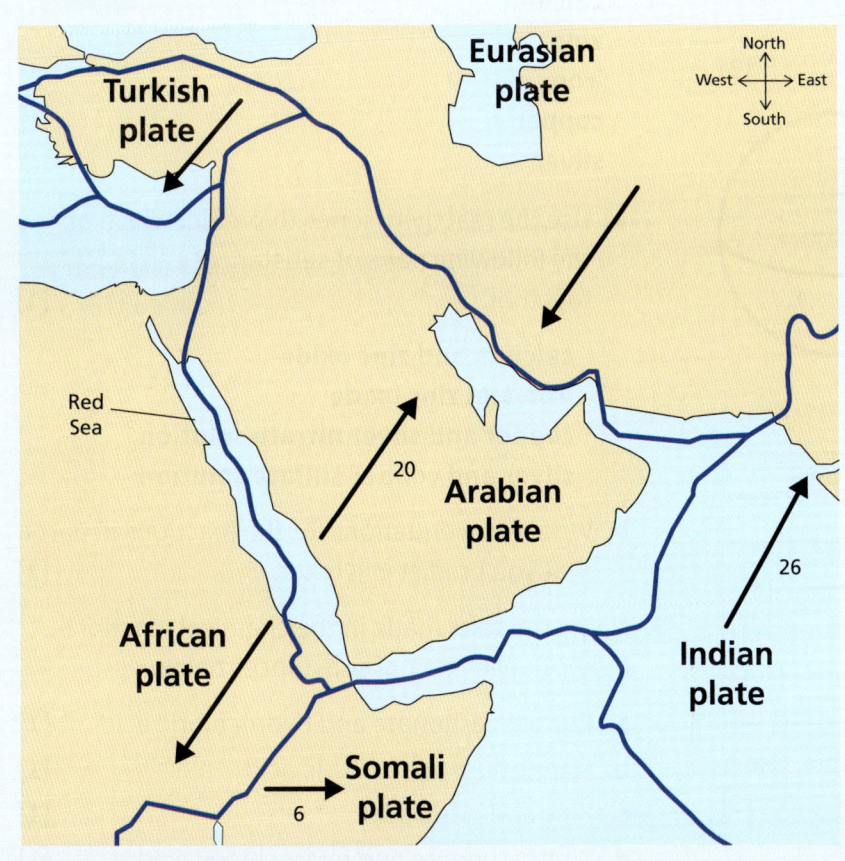

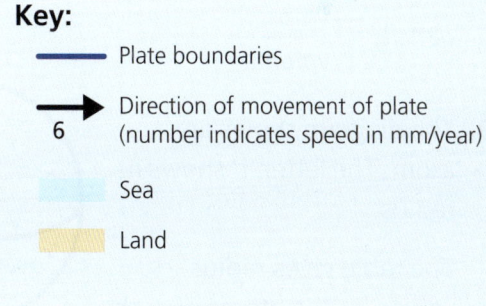

Stage 9 Review

1. A student draws the electron configurations of four atoms.

   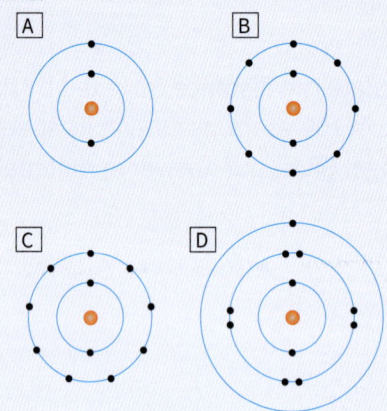

   a. Which electron arrangement is incorrect? [1]
   b. Which two elements are in the same group of the periodic table? [1]
   c. Which element has a proton number of 10? [1]

2. A student has a block of chromium. Its volume is 4 cm³. It has a mass of 28 g. Use the equation below to calculate its density.

   $$\text{density} = \frac{\text{mass}}{\text{volume}}$$ [3]

3. The diagram shows an atom. The letter 'r' shows its radius.

   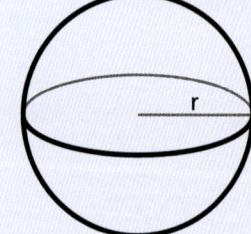

   The table gives radius values for the atoms of each element in Group 1 of the periodic table.

Element	Atomic radius (pm)
lithium	152
sodium	186
potassium	231
rubidium	244
caesium	262

   a. Draw a bar chart for the data in the table. Use a copy of the axes above right. [3]

   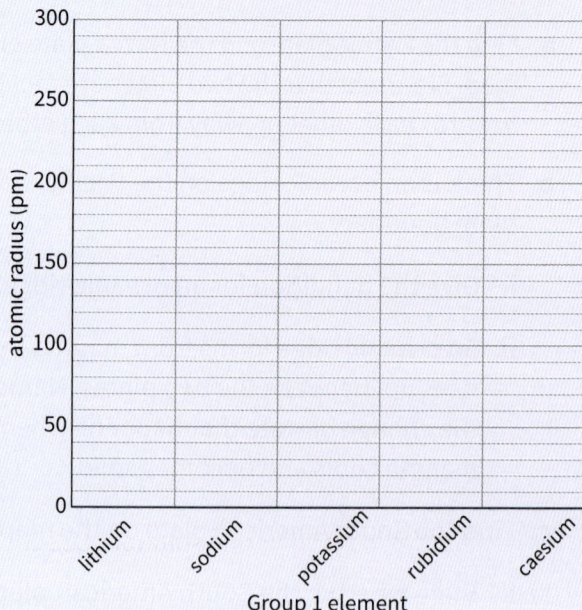

   b. Describe the pattern shown on the graph. [1]
   c. The radius of a sodium ion, Na⁺, is 95 pm.
      i. Compare the radius of a sodium ion to a sodium atom. [1]
      ii. Suggest a reason for the difference in radius of a sodium ion and a sodium atom. In your answer, include diagrams of the electron configurations of the atom and ion. [3]

4. Part of the reactivity series is given below.

   **calcium
   zinc
   iron
   copper
   silver**

   a. Use the reactivity series to predict which of the following pairs of substances can react together. [1]

   **calcium and zinc oxide
   iron and zinc oxide
   copper and silver nitrate solution
   silver and copper sulfate solution**

   b. Write word equations for the reactions in part **a** that you predict will react. [2]

5. Name the salts made in the chemical reactions between each pair of substances below.
   a. Copper carbonate and hydrochloric acid [1]
   b. Magnesium and sulfuric acid [1]
   c. Zinc and nitric acid [1]
   d. Zinc carbonate and hydrochloric acid [1]

Stage 9 review

6. Copy and complete the word equation below.
   a. magnesium + nitric acid → [1]
   b. zinc + sulfuric acid → [1]
   c. magnesium + hydrochloric acid → [1]
   d. copper carbonate + hydrochloric acid → [1]

7. Zara wants to make zinc chloride. She adds zinc to hydrochloric acid, until a little zinc remains unreacted. Then she filters the mixture.
   a. Explain why she filters the mixture. [1]
   b. Describe how Zara can make zinc chloride crystals from zinc chloride solution. Use the words *evaporation* and *crystallisation* in your answer. [4]

8. A scientist investigates the question *How does temperature affect the speed of rusting?* She places six identical iron nails in boiling tubes. The nails are exposed to both air and water. She places each boiling tube in an oven or fridge at a different temperature. She observes the nails regularly.

   Her results are in the table.

Temperature (°C)	Time for rust to appear on iron nail (hours)
10	240
20	120
30	100
40	30
50	15
60	7

   a. Finish labelling the axes on a copy of the graph axes below. [2]
   b. Plot the data in the table on your graph. [2]
   c. Draw a line of best fit. [1]
   d. Draw a circle around the anomalous result. [1]

   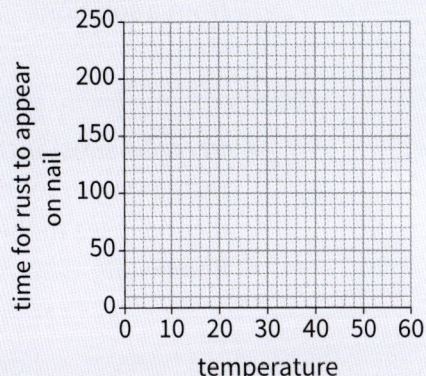

9. Abbas investigates the reaction of zinc with hydrochloric acid. He pours 25 cm³ of the acid into a conical flask. He adds small pieces of zinc. He collects the gas in a syringe.

   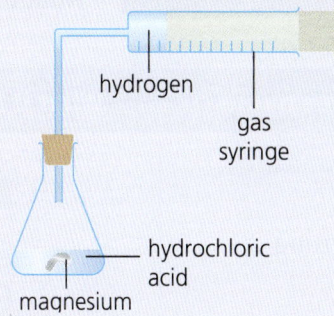

   a. The products of the reaction are zinc chloride and hydrogen. Write a word equation for the reaction. [1]
   b. Abbas measures the volume of gas collected every minute. He plots a graph of his results.

   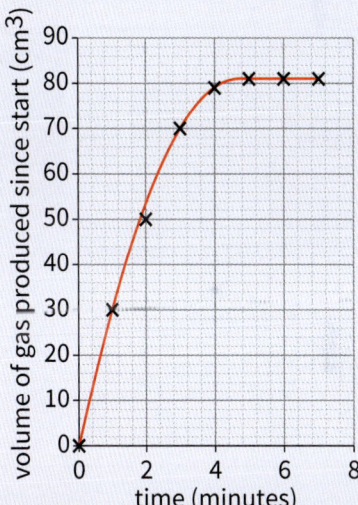

   i. Between which times is the reaction fastest? Choose from the list below. [1]

   Between 0 and 1 minute. Between 2 and 3 minutes. Between 3 and 4 minutes.

   ii. After how many minutes does the reaction finish? [1]
   iii. What is the total volume of gas made in the investigation? [1]

   c. Abbas wants to find out how the rate of reaction changes if he increases the concentration of acid.
   i. Name two variables he should keep the same in his investigation. [2]
   ii. Explain why he should keep these variables the same. [1]

237

Reference 1

Choosing apparatus

There are many different types of scientific apparatus. The table below shows what they look like, how to draw them, and what you can use them for.

Apparatus name	What it looks like	Diagram	What you can use it for
test tube			• heating solids and liquids • mixing substances • small-scale chemical reactions
boiling tube			• a boiling tube is a big test tube; you can use it for doing the same things as a test tube
beaker			• heating liquids and solutions • mixing substances
conical flask			• heating liquids and solutions • mixing substances
filter funnel			• to separate solids from liquids, using filter paper
evaporating dish			• to evaporate a liquid from a solution
condenser			• to cool a substance in the gas state, so that it condenses to the liquid state

Reference

stand, clamp, and boss			• to hold apparatus safely in place
Bunsen burner			• to heat the contents of beakers or test tubes • to heat solids
tripod			• to support apparatus above a Bunsen burner
gauze			• to spread out thermal energy from a Bunsen burner • to support apparatus, such as beakers, over a Bunsen burner
pipette			• to transfer liquids or solutions from one container to another
syringe			• to transfer liquids and solutions • to measure volumes of liquids or solutions
spatula			• to transfer solids from one container to another
tongs and test tube holders			• to hold hot apparatus, or to hold a test tube in a hot flame

Reference 2

Working accurately and safely

You need to make accurate measurements in science practicals. You will need to choose the correct measuring instrument, and use it properly.

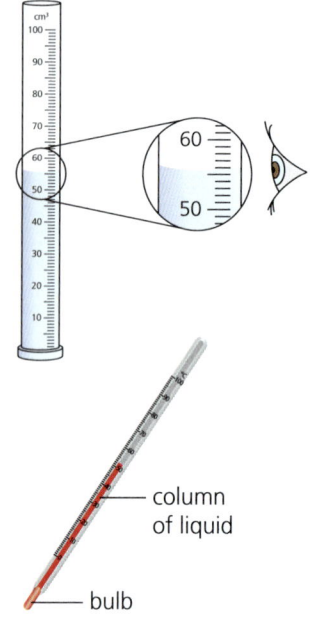

Measuring cylinder

Measuring cylinders measure volumes of liquids or solutions. A measuring cylinder is better for this job than a beaker because it measures smaller differences in volume.

To measure volume:

1. Place the measuring cylinder on a flat surface.

2. Bend down so that your eyes are level with the surface of liquid.

3. Use the scale to read the volume. You need to look at the bottom of the curved surface of the liquid. The curved surface is called the **meniscus**.

Measuring cylinders measure volume in cubic centimetres, cm^3, or millilitres, ml. One cm^3 is the same as one ml.

▲ The different parts of a thermometer.

Thermometer

The diagram to the left shows an alcohol thermometer. The liquid expands when the bulb is in a hot liquid and moves up the column. The liquid contracts when the bulb is in a cold liquid.

To measure temperature:

1. Look at the scale on the thermometer. Work out the temperature difference represented by each small division.

2. Place the bulb of the thermometer in the liquid.

3. Bend down so that your eyes are level with the liquid in the thermometer.

4. Use the scale to read the temperature.

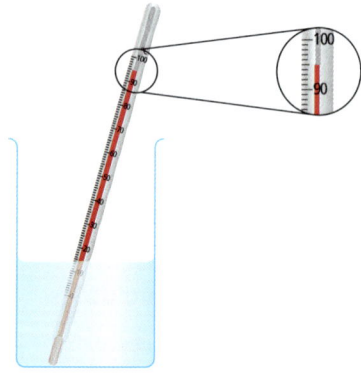

▲ The temperature of the liquid is 95 °C.

Most thermometers measure temperature in degrees Celsius, °C.

Balance

A **balance** is used to measure mass. Sometimes you need to find the mass of something that you can only measure in a container, like liquid in a beaker. To use a balance to find the mass of liquid in a beaker:

1. Place the empty beaker on the pan. Read its mass.

2. Pour the liquid into the beaker. Read the new mass.

3. Calculate the mass of the liquid like this:

   (mass of liquid) = (mass of beaker + liquid) − (mass of beaker)

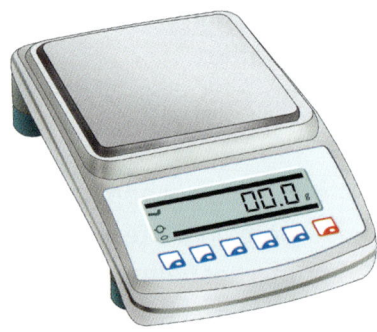

▲ The balance measures mass.

Balances normally measure mass in grams, g, or kilograms, kg.

Working safely

Hazard symbols

Hazards are the possible dangers linked to using substances or doing experiments. Hazardous substances display **hazard symbols**. The table shows some hazard symbols. It also shows how to reduce risks from each hazard.

Hazard symbol	What it means	Reduce risks from this hazard by…
	Corrosive – the substance attacks and destroys living tissue, such as skin and eyes.	• wearing eye protection • avoiding contact with the skin
	Irritant – the substance is not corrosive, but will make the skin go red or form blisters.	• wearing eye protection • avoiding contact with the skin
	Toxic – can cause death, for example, if it is swallowed or breathed in.	• wearing eye protection • wearing gloves • wearing a mask, or using the substance in a fume cupboard
	Flammable – catches fire easily.	• wearing eye protection • keeping away from flames and sparks
	Explosive – the substance may explode if it comes into contact with a flame or heat.	• wearing eye protection • keeping away from flames and sparks
	Dangerous to the environment – the substance may pollute the environment.	• taking care with disposal

Other hazards

The table does not list all the hazards of doing practical work in science. You need to follow the guidance below to work safely. Always follow your teacher's safety advice, too.

- Take care not to touch hot apparatus, even if it does not look hot.
- Take care not to break glass apparatus – leave it in a safe place on the table, where it cannot roll off.
- Support apparatus safely. For example, you might need to weigh down a clamp stand if you are hanging heavy loads from the clamp.
- If you are using an electrical circuit, switch it off before making any change to the circuit.
- Remember that wires may get hot, even with a low voltage.
- Never connect wires across the terminals of a battery.
- Do not look directly at the Sun, or at a laser beam.
- Wear eye protection – whatever you are doing in the laboratory!

Glossary

Accurate Accurate data is data that is close to the true value.

Acidic A solution is acidic if its pH is less than 7.

Acidity A chemical property that describes how acidic a substance is.

Alkaline A solution is alkaline if its pH is greater than 7.

Alkalinity A chemical property that describes how alkaline a substance is.

Alloy A mixture of a metal with one or more other elements.

Analogy A comparison between one thing and another that helps to explain something. It can be used as a model.

Atom The smallest part of an element that can exist.

Atomic number The number of protons in an atom of an element. Also called the proton number.

Balancing number The balancing numbers in a chemical equation show the ratio of the numbers of particles of the reactants and products. A balancing number is written to the left of its chemical formula.

Boiling The change of state from liquid to gas that only happens if the liquid is hot enough.

Boiling point The temperature at which a substance boils.

Brittle A material is brittle if it breaks easily when hit with a hammer.

Changes of state The change that happens when a substance changes from one state to another.

Chemical formula The chemical formula of a substance gives the relative number of atoms of each element in the substance.

Chemical properties Properties that describe how substances change in chemical reactions.

Chemical reaction A change in which atoms are rearranged and join together differently to make new substances.

Chemical symbol The internationally recognised one- or two-letter code for an element.

Chromatography A method to separate and identify the substances in a mixture. It works if all the substances in a mixture dissolve in the same solvent.

Collaborate To work together.

Combustion reaction A burning reaction, in which a substance reacts quickly with oxygen and gives out light and heat.

Compound A substance made up of atoms of two or more substances, strongly joined together.

Compress Make smaller by pressing.

Concentrated A solution containing a large amount of solute dissolved in a solvent.

Concentration The amount of solute that is dissolved in a certain volume of solution or solvent.

Condensing or condensation The change of state from gas to liquid.

Conductor A substance that allows heat and/or electricity to pass through it easily.

Continental drift The movement of tectonic plates over millions of years.

Corrosion A chemical reaction that happens slowly on the surface of a metal.

Covalent bond A shared pair of electrons that holds two atoms together.

Crust The outer layer of the Earth, made of solid rock.

Density The mass of a material in a certain volume. It is calculated using the formula $density = \frac{mass}{volume}$.

Desalination The process of removing salt from seawater.

Dilute A solution containing a small amount of solute dissolved in a large volume of solvent.

Ductile A material is ductile if it can be pulled into wires.

Electron A tiny sub-atomic particle with a negative charge that moves around in an atom, outside the nucleus. It has a single negative charge. Its relative mass is $\frac{1}{2000}$.

Electron configuration The arrangement of electrons in an atom or ion.

Element A substance that is made of one type of atom that cannot be split into other substances.

Glossary

Endothermic change A physical or chemical change that transfers energy from the surroundings to the reaction mixture.

Energy conservation Energy is never made or lost but can be transferred from one form to another, although this is not always a form that can be used.

Evaporating or evaporation The change of state from liquid to gas that can happen at any temperature.

Exothermic change A physical or chemical change that transfers energy to the surroundings from the reaction mixture.

Flammable If a material burns easily, it is flammable.

Fold mountains Mountains that form when tectonic plates push together.

Freezing The change of state from liquid to solid.

Giant covalent structure A three-dimensional network of atoms that are joined together by covalent bonds.

Giant ionic structure The three-dimensional structure of positive and negative ions in an ionic compound.

Giant metallic structure A three-dimensional pattern of positive metal ions held together by moving electrons.

Group 1 The elements in the left column of the periodic table – lithium, sodium, potassium, rubidium, caesium.

Group 2 The elements in the second column of the periodic table – beryllium, magnesium, calcium, strontium, barium.

Hazard A possible source of danger.

Hazard symbol A warning symbol on a substance that shows what harm it might cause if not handled properly.

Indicator A solution of a dye that turns a different colour in acidic and alkaline solutions.

Inert A substance is inert if it does not take part in chemical reactions.

Inner core The solid iron and nickel at the centre of the Earth.

Insoluble A substance is insoluble in a solvent if it cannot dissolve in the solvent.

Insulator A substance that does not conduct electricity is an insulator.

Ion A particle with a positive or negative charge, formed when an atom loses or gains electrons.

Ionic bonding The electrostatic attraction between positive and negative ions in a giant ionic structure. Ionic bonds act in all directions.

Ionic compound A compound made up of ions. Most compounds made up of a metal and a non-metal are ionic.

Isotopes Atoms of the same element that have different numbers of neutrons are called isotopes.

Lava Liquid rock that is on, or above, the surface of the Earth.

Malleable A material is malleable if it can be hammered into shape without cracking.

Mantle The layer of the Earth beneath the crust. It is solid but can flow slowly. It goes down almost halfway to the centre of the Earth.

Mass conservation Mass is conserved when the total mass of the substances before a change is equal to the total mass of the substances after the change. Mass is conserved in changes such as chemical reactions, dissolving, and changes of state.

Mass number The total number of protons and neutrons in an atom.

Materials The different types of matter that things are made from.

Melting The change of state from solid to liquid.

Melting point The temperature at which a substance changes from the solid to the liquid state.

Metal An element on the left of the stepped line of the periodic table. Most metals are good conductors of heat and electricity.

Metal displacement reaction In a metal displacement reaction, a more reactive metal displaces (pushes out) a less reactive metal from its compound.

Mineral salts/minerals Dissolved compounds needed by plant and animal cells to grow and remain healthy.

Mixture A mixture contains two or more substances that are not joined together.

Model A way of representing something that you cannot see or experience directly. A model may be a physical model built on a different scale to the original system, or it may take the form of equations.

Glossary

Model of plate tectonics The model that suggests that the Earth's crust and uppermost mantle are made of about 12 slabs of solid rock. It can explain earthquakes, volcanoes and the formation of some mountains.

Molecule A particle made up of two or more atoms, strongly joined together.

Neutral (1) A solution is neutral if its pH is 7 – for example, pure water is neutral. (2) A particle is neutral if it is neither positively nor negatively charged, for example an atom of an element or a neutron in the nucleus of an atom.

Neutralisation A type of chemical reaction in which an alkali reacts with an acid and the pH gets closer to 7.

Neutron A tiny sub-atomic particle with no charge that is found in the nucleus of an atom. The relative mass of a neutron is 1.

Non-metal An element on the right of the stepped line of the periodic table. Most non-metals do not conduct heat or electricity.

Nucleon number The total number of protons and neutrons in an atom.

Nucleons Protons and neutrons.

Nucleus The central part of an atom, made up of protons and neutrons.

Oceanic ridge A mountain chain on the sea floor.

Ore A rock that a metal can be extracted from.

Outer core The liquid iron and nickel between the Earth's mantle and inner core.

Particle model This describes the arrangement and movement of particles in a substance.

Particles The tiny pieces of matter that everything is made from.

Peer review The checking of scientific research by other experts.

Physical properties Properties that you can observe or measure without changing a material.

Precipitate A suspension of tiny solid particles mixed with a liquid or solution.

Precipitation reaction A chemical reaction in which two reactants in solution react together to make a precipitate.

Products The substances that are made in a chemical reaction.

Properties The properties of a substance describe what it is like and what it does.

Proton A tiny sub-atomic particle with a positive charge that is found in the nucleus of an atom. The relative mass of a proton is 1.

Proton number The number of protons in an atom of an element. Also called the atomic number.

Pure substance A substance that consists of one substance only. It is not mixed with anything, and all its particles are identical.

Purity This describes how much of a substance is in a mixture.

Rate of a reaction The rate of a chemical reaction is the amount of reactant used up, or the amount of product made, in a given time.

Reactants The starting substances in a chemical reaction.

Reactivity series A list of metals in order of how readily they react with other substances, such as water, oxygen, and dilute acids.

Reactivity The tendency of a substance to take part in a chemical reaction.

Reliable You can obtain reliable data by making enough measurements.

Risk The chance of injury from a hazard. A combination of the probability that something will happen and the consequence if it did.

Salt A compound made when a metal ion replaces the hydrogen ion in an acid.

Saturated solution A solution in which no more solute can dissolve. The solution is in contact with undissolved solute.

Scientific question A question that can be answered using evidence or data.

Seafloor spreading The movement of the seafloor away from the two sides of an oceanic ridge.

Secondary data Evidence or data that has been collected by someone else.

Solubility The maximum mass of solute that can dissolve in 100 g of solvent.

Soluble A substance is soluble in a solvent if it can dissolve in the solvent.

Glossary

Solute The substance that dissolves in a solvent to make a solution.

Solution A mixture that forms when a substance dissolves in a liquid.

Solvent In a solution, the liquid in which the solute is dissolved.

Sonorous Metals are sonorous – they make a ringing sound when hit.

States of matter Most substances can exist as a solid, liquid, or gas. These are the states of matter.

Steels Alloys of iron.

Strong A material is strong if a large force is needed to break it.

Sub-atomic particles The particles that make up an atom, including protons and neutrons in the nucleus, and electrons outside the nucleus.

Substance A material that has one type of matter.

Symbol equation A symbol equation uses chemical formulae to represent a chemical reaction. It shows the relative number of particles of each of the reactants and products.

Systematic review A systematic review uses repeatable methods to collect and analyse secondary data from many scientists.

Tectonic plates The 12 or so slabs of solid rock that make up the Earth's crust and uppermost mantle.

Universal indicator A mixture of dyes that changes colour to show how acidic or alkaline a solution is.

Vacuum A space that has no particles, and so no matter.

Vitamins Substances needed in tiny amounts in the diet to help chemical reactions take place in cells.

Volcano An opening in the Earth's crust that liquid rock and other materials escape from.

Word equation A word equation summarises a chemical reaction in words. It shows the reactants and products. The arrow means 'react to make'.

Index

accuracy 135
acidity 88–90
acid rain 96–97, 147, 149
acid reactions 77, 88–101, 149–151, 159–165, 198–199, 205, 211–217, 222–225
acid rivers 161
acid spills 133
al-Biruni, Abu Rayhan 192–193
alkalinity 88–89
alkalis 88–101
alloys 66–67, 70–73, 84–85
aluminium 209
ammonia 178
analogy 114–115, 117, 190–191
animal fossils 104, 231
appearance, metals 63
arrangement
　electrons 172–179, 182–183
　elements 44–45
　particles 21–26, 30–31
atomic mass 44
atomic number 170
　see also proton number
atoms
　chemical reactions 78–83
　definitions 42–43
　modelling 43, 116–119
　physical properties 67–71
　states of matter 114–125
　structure 114–123, 170–185, 195
attraction 21–22, 115

balancing numbers 203–205
bamboo bikes 73
bar charts 142–143
bicycles 72–73
body substances 58–59
boiling 27–29
boiling points 27–29, 31, 69, 181, 195–196
bonding 176–185, 195
Bose–Einstein condensate 120
Bose, Satyendra Nath 120
Brandt, Hennig 38
brittle materials 68–69, 71, 73, 177
bronze 38, 67

calcium 158, 160
calculating density 188–189
Cannizzaro, Stanislao 44
carbon 208–209
carbonates 98–99, 215–217
carbon dioxide 47, 98
carbon fibre reinforced polymers (CFRP) 73
carbon monoxide 47
cathode rays 116–117
Cavendish, Henry 39
CFRP see carbon fibre reinforced polymers
changes of state 26–31, 151
charge, mass 115
chemical formulae 50–51, 175
chemical properties 62, 67, 82–89, 172–185, 188–201
chemical reactions 76–85, 88–101, 114, 136–137, 146–169, 174–175, 196–227
　acids 77, 88–101, 149–151, 159–165, 198–199, 211–217, 222–225
　energy 150–155, 202–203
　mass 79–81, 202–203
　metals 90–95, 98–99, 156–165, 196–199, 204–217, 222–225
　oxygen 146–147, 156–157, 204–207, 215
　products 77–83, 92, 95–99, 146–149, 156–163, 202–207, 211–218
　reactants 77–83, 92, 96, 146–150, 202–204, 212–218, 221
　states of matter 76–87
　word equations 148–149, 156–158, 161, 203–207
chemical symbols 40–41
chlorine 136–137
chromatography 130–131
citric acid 92, 150
classifying acids and alkalis 90–91
coatings 85
collaborate/collaborating 109, 120–121
combustion reactions 82, 84, 149, 156, 202–203, 206
compounds 46–53, 56–59, 176–177, 179, 183–185
compress 25
compression 25
concentration 132–133, 137, 220–221
conclusions 29–31, 64–65, 81, 95, 143, 155, 165, 181
condensing/condensation 27
conductors 63–65
conserving energy 202–203
conserving mass 202
continental drift 104, 228–233
continuous variable 139, 143, 219, 223
control variables 64–65, 152, 164
copper 38, 67, 156–157, 160–162, 210–211
copper sulfates 216–217
corrosion reactions 82–83, 149, 156, 206
covalent bonding/structures 178–181, 183–185
crust 102–103
crystallisation 213, 217

Dalton, John 44
data presentation 28–31, 90, 135–143, 153, 195–198, 218–223
dehydration 184–185
density 72, 188–193, 196–197, 209
dependent variables 64–65, 141–143, 152, 164
desalination 128–129
diarrhoea 184–185
dilute 132–133
discovering electrons 116–117
discovering the elements 38–39
displacement reactions 206–207, 211
dissolving 54–55, 76, 82–83, 96–97, 128–143, 151, 160–161, 185, 221
dot-and-cross diagrams 178–179
drinking seawater 128–129
ductile, metals 63

Earth 102–113, 228–235

Index

earthquakes 103, 106–107
Einstein, Albert 120
electric charge 115
electricity 63, 69, 209
electron configuration 172–185, 195
electrons 114–123, 172–185, 195–199, 206
elements 36–75, 79, 82–83
endothermic reactions 150–151, 203
energy 150–155, 202–203
environmental aspects 147, 210–211
evaporating/evaporation 26–27, 213
evidence analysis
    chemical reactions 95, 153–155, 164–165, 216–225
    Earth 102–105, 230–231
    particle model 28–29
    states of matter 114–121, 135–138, 141–143
exothermic reactions 150–151, 202–203
extracting metals 38, 208–211

fair tests 64–65, 94, 134, 140, 152, 164, 220, 222
filtration 213, 217
flammable 76–77, 213
flow 24–25
fold mountains 107
food energy 154–155
formulae 50–51, 175
fossils 104, 230–232
freezing 26, 30–31, 33

gas products 98–99
gas state 22–31, 69, 188–189
Geiger, Hans 118–119
gemstone densities 192–193
giant covalent structures 180–181, 183
giant ionic structures 176–177, 183–185
giant metallic structures 182–183, 195
glossopteris trees 230–231

gold 156–157, 160, 209
graphs 127, 142–143
Group 1 elements 194–197
Group 2 elements 198–199

Hayyan, Jabir Ibn 91
hazards 134–135
hazard symbols 77, 82–83, 164, 213
health, nitrogen dioxide 147
heat sinks 64–65
Higgs Boson 120–121
Higgs, Peter 120–121
history, elements 38–39
Hittorf, Johann 116
hot spots 108–109
hydrochloric acid reactions 77, 90–95, 98–99, 150, 160–165, 198–199, 205, 211–218, 222–225
hydrogen cars/fuels 78
hydrogen tests 99
hypothesis 28–29, 80–81, 104–105, 120–121, 154–155, 164–165, 220–223

ice 22–27
independent variables 64–65, 141–143, 152, 164
indicators 88–95, 158
inert metals 162
inks 130
inner core 102–103
inside atoms 114–125
insoluble/insolubility 54–55, 82–83, 216–217, 222
insulators 68–69
interviews 137
investigations
    atoms 116–119
    chemical reactions 80–83, 94–95, 152–155, 164–165, 220–225
    chromatography 130–131
    electrons 116–119
    oral rehydration solutions 185
    physical properties 64–65
    substance purity 127, 131–143
iodine deficiency 59
ionic bonding 176–177, 179, 183–185

ionic compounds 176–177, 179, 183–185
ions 174–177, 182–185, 195, 206, 212
iron 159–160
iron alloys 66–67, 70–73
isotopes 123

Jabir Ibn Hayyan 91

Large Hadron Collider (LHC) 120–121
lava 108–109
lead, extraction 208
Lehmann, Inge 103
LHC see Large Hadron Collider
life-saving compounds 184–185
line graphs 142–143
liquid state 22–31, 188–189
lystrosaurus 231

magnesium chloride 212–213
magnesium reactions 76–77, 79, 156, 158, 160, 204–205
magnetic stripes 233
malleable, metals 63
mantle 102–103
Marsden, Ernest 118–119
mass
    charge 115
    chemical reactions 79–81, 202–203
    conservation 202
    density 189–190
    mass number 122–123
    measurements 188–189
materials
    elements 36
    particle model 20–23
    physical properties 22–25, 62–75
matter
    particle model 20–31
    states of 22–31, 36–37, 76–87, 114–147
measuring mass 188–189
measuring pH 91
medical oxygen 68
melting 26, 30–31, 33, 43

Index

melting points 30–31, 33, 70–71, 194–196, 198–199
Mendeleev, Dmitri 44–45
mesosaurus 104, 231
metallic bonding/structures 182–183, 195
metal oxides 206–207
metals
    acid reactions 90–95, 98–99, 160–165, 198–199, 205, 211–217, 222–225
    chemical reactions 90–95, 98–99, 156–165, 196–199, 204–217, 222–225
    compounds 48–49
    displacement reactions 206–207, 211
    extraction 38, 208–211
    Group 1 194–197
    oxygen reactions 156–157, 204–207, 215
    periodic table 36–39, 48–49, 194–197
    physical properties 62–73
    reactivity series 162–165
    salt formation 212–215
    water reactions 158–159, 196–197
minerals 59
mines 103
mixtures 52–59, 66–67, 70–73, 156–157, 213, 217
modelling
    atoms 43, 114–119
    concentration 132
    density 191
    Earth 102–105
    ionic bonding 177
    particle model 20–35
    in science 32–33
molecules 46–54, 70–71, 78, 126, 180–181
mountain formation 107
movement, particles 21–26, 30–31

naming compounds 48–49
neutralisation reactions 92–98, 150–151
neutral substances 88–98
neutrons 114–115, 119–123, 170–174
nitrogen 146–147, 180
nitrogen dioxide 147
nitrogen monoxide 146–147
Noddack, Walter 45
non-metals 36–39, 48–49, 62, 66–73, 178–184
nucleon numbers 122–123
nucleons 122–123
    see also neutrons; protons
nucleus 114–115, 118–123, 170–174, 195–199

oceanic ridges 232–233
oil coatings 85
oral rehydration salts 184–185
order of reactivity 157, 159, 161–165
ores 208–211
organising elements 44–45
outer core 102–103
oxygen
    chemical reactions 76–85, 146–149, 156–159, 202–209
    compounds 46–52
    gas tests 99
    medical purposes 68

packing, particles 190–191
paint coatings 85
particles
    elements 36, 42–43, 46–57
    particle diagrams 52–53, 206–207
    particle mass 190
    particle model 20–35
    particle packing 190–191
    physical properties 67, 70–71
patterns, periodic table 188–201
peer review 104–105, 121
periodic table 36–41, 44–45, 48, 62, 68, 122, 170–173, 183, 188–201
phosphorous 38
pH scale 90–95
physical properties 22–25, 62–75, 172–185, 196–199
phytomining 211
planet Earth 102–113, 228–235
plate tectonics 104–109, 228–234
plum pudding model 117–119

potassium 158, 160
precipitates 82–83
precipitation reactions 82–83, 148, 204, 206, 222
precision 193
predictions 65, 80–81, 106–107, 109, 141
preventing corrosion 85
Priestley, Joseph 39
products 77–83, 92, 95–99, 146–149, 156–163, 202–207, 211–218
properties 170–187
    alloys 66–67, 70–73
    atoms 43
    chemical properties 62, 67, 82–89, 172–185, 188–201
    compounds 46–52, 56–57
    elements 36–38, 42–57
    metals 62–73
    non-metals 62, 66, 68–71, 73
    particle model 20–31
    physical properties 22–25, 62–75, 172–177, 180–183, 196–199
proton numbers 122–123, 170–171
protons 114–115, 119–123, 170–174
pure/purity, substances 126–145

quartz, density 192

rate of reaction 218–225
reactants 77–83, 92, 96, 146–150, 202–204, 212–218, 221
reaction rates 218–225
reactions *see* chemical reactions
reactivity 156–165, 199, 206–211
reactivity series 162–165, 206–211
recycling 211
rehydration salts 184–185
reliable data 135
report writing 137, 142–143, 155
restless Earth 106–107
risk assessments 134–135, 152, 164, 213
rivers, acidity 161
Rutherford, Daniel 39
Rutherford, Ernest 118–119

Index

safety 81–83, 94–95
salts 134–135, 212–217
saturated solution 138–139
Scott, Robert Falcon 230
seafloor spreading 232–233
seawater 126, 128–129, 134–135
secondary data 90, 138–139
seismometer 107
separation, particles 21–26, 30–31
shape
  metals 63
  particle model 24–25
silver 160
simple covalent structures 180–181, 183
sodium 156–158, 160
sodium chloride 174–176
sodium hydrogencarbonate 82, 92, 150
soil pH 93
solid state 22–31, 69, 188–189
soluble/solubility 54–55, 82–83, 138–143, 216–217, 222
solutes 54–55, 132–133, 138
solutions 54–55, 132–134, 138–139, 148–151, 158–160, 207
solvents 54–55, 130, 132
sonorous/sonority 63
stability, ions 175
stainless steel 85
state, metals 63
states of matter 22–31, 36–37, 76–87, 114–147, 151, 188–189

state symbols 204–205
steam 22–27
steel 66–67, 70–73
strength
  metals 63, 71
  non-metals 71
structure 70, 195, 206
  atoms 114–123, 170–185, 195
  bonding and properties 170–187
  Earth 102–103
sub-atomic particles 114–123, 170–174
  see also electrons
substances
  elements 36–38, 42–47, 50–59
  particle model 20–27, 30–33
  pure/purity 126–145
sulfuric acid 91, 133, 211–212, 216–217
surface area 224–225
sweets 130
symbols
  chemical symbols 40–41
  hazard symbols 77, 82–83, 164, 213
  state symbols 204–205
  symbol equations 203–205
systematic review 184–185

Tacke, Ida 39, 45
tectonic plates 104–109, 228–234

temperature 127, 139–143, 150–151, 222–223
thermometers 28–29, 141
Thomson, Joseph John 116–119
Tinto River, Spain 161
tree fossil records 230–231
twentieth and twenty-first century discoveries 39

universal indicators 91–95, 158

vacuum 24–25
variables
  chemical reactions 94, 152, 164–165
  physical properties 64–65
  solubility 140–143
vitamins 58–59
volcanoes 108–109
wastes, extracting metals 211
water
  chlorine 136–137
  metal reactions 158–159, 196–197
  properties 46–47
  states of matter 22–27
Wegener, Alfred 104–105, 231–232
word equations 148–149, 156–158, 161, 203–207

zinc 38, 85, 156–157, 159, 205
zinc sulphate 215

OXFORD
UNIVERSITY PRESS

Great Clarendon Street, Oxford, OX2 6DP, United Kingdom

Oxford University Press is a department of the University of Oxford. It furthers the University's objective of excellence in research, scholarship, and education by publishing worldwide. Oxford is a registered trade mark of Oxford University Press in the UK and in certain other countries

© Oxford University Press 2021

The moral rights of the author have been asserted

First published in 2021

All rights reserved. No part of this publication may be reproduced, stored in a retrieval system, or transmitted, in any form or by any means, without the prior permission in writing of Oxford University Press, or as expressly permitted by law, by licence or under terms agreed with the appropriate reprographics rights organization. Enquiries concerning reproduction outside the scope of the above should be sent to the Rights Department, Oxford University Press, at the address above.

You must not circulate this work in any other form and you must impose this same condition on any acquirer

British Library Cataloguing in Publication Data

Data available

978-1-38-201848-7

Digital edition: 978-1-38-201854-8

10 9 8 7 6 5

Paper used in the production of this book is a natural, recyclable product made from wood grown in sustainable forests. The manufacturing process conforms to the environmental regulations of the country of origin.

Printed in China by Golden Cup

Acknowledgements

The publisher and authors would like to thank the following for permission to use photographs and other copyright material:

Cover: Dario Bosi/Getty Images. Photos: p18: Arcady/Shutterstock; p20(t): Roop_Dey/Shutterstock; p20(bl): Olaf Speier/Shutterstock; p20(bm): VitaminCo/Shutterstock; p20(br): Salineekapui/Shutterstock; p21(l): Inc/Shutterstock; p21(m): George Dolgikh/Shutterstock; p21(m): Stephen E Bishop/Shutterstock; p21(r): Mark S Johnson/Shutterstock; p22(tl): Valentyn Volkov/Shutterstock; p22(tm): kubais/Shutterstock; p22(tr): Phuangphet geissler/Shutterstock; p22(m): luchschenF/Shutterstock; p22(b): icedmocha/Shutterstock; p24(t): Ivo Petkov/Shutterstock; p24(m): Elzbieta Krzysztof/Shutterstock; p24(b): myboys.me/Shutterstock; p25(tl): MarcelClemens/Shutterstock; p25(tr): sruilk/Shutterstock; p25(b): ChiccoDodiFC/Shutterstock; p26(t): EtiAmmos/Shutterstock; p26(m): SGM/Shutterstock; p26(b): Teim/Shutterstock; p28(tl): KishoreJ/Shutterstock; p28(tr): suchitra poungkoson/Shutterstock; p28(bl): trambler58/Shutterstock; p28(bm): withGod/Shutterstock; p28(br): Olga Popova/Shutterstock; p30(t): NataliSel/Shutterstock; p30(m): megaflopp/Shutterstock; p30(b): Ieva Zigg/Shutterstock; p31: Martyn F. Chillmaid/Science Photo Library; p32(tl): Ekaterina_Minaeva/Shutterstock; p32(tr): Art Konovalov/Shutterstock; p32(b): Tugce Simsek/Shutterstock; p33(tr): Arsel Ozgurdal/Shutterstock; p33(ml): Pyty/Shutterstock; p33(mr): StudioMolekuul/Shutterstock; p34: StefanRenner/Shutterstock; p36(t): Mr.Jakrapong phoaphom/Shutterstock; p36(bl): Photo Oz/Shutterstock; p36(bm): Science Photo Library; p36(br): www.BibleLandPictures.com/Alamy Stock Photo; p37(tl): Primož Cigler/Shutterstock; p37(tm): Happy monkey/Shutterstock; p37(tr): Elenamiv/Shutterstock; p37(bl): Volodymyr Krasyuk/Shutterstock; p37(br): Pearl-diver/Shutterstock; p38(tl): www.BibleLandPictures.com/Alamy Stock Photo; p38(tm): Khoroshunova Olga/Shutterstock; p38(tr): SOMMAI/Shutterstock; p38(mr): robertharding/Alamy Stock Photo; p38(ml): Dnaniss/Shutterstock; p38(b): Ashmolean Museum/Bridgeman Images; p40: Pixel-Shot/Shutterstock; p41: Peter Hermes Furian/Shutterstock; p42(tl): bonchan/Shutterstock; p42(tr): Karynav/Shutterstock; p42(b): Laurence Marks, Northwestern University/Science Photo Library; p43: Dmytro Falkowskyi/Shutterstock; p44(tl): Andrew Lambert Photography/Science Photo Library; p44(tm): Oksana2010/Shutterstock; p44(tr): Papa1266/Shutterstock; p44(b): Sputnik/Science Photo Library; p45(t): Bjoern Wylezich/Shutterstock; p45(b): Science History Images/Alamy Stock Photo; p46(l): Jason Stitt/Shutterstock; p46(m): Andrew Lambert Photography/Science Photo Library; p46(r): Charles D. Winters/Science Photo Library; p48(tl): Fablok/Shutterstock; p48(tr): Jirik V/Shutterstock; p48(b): Miro Novak/Shutterstock; p49(t): Baloncici/Shutterstock; p49(ml): DKN0049/Shutterstock; p49(mr): tersetki/Shutterstock; p49(b): underdog_cg/Shutterstock; p50(t): Maderla/Shutterstock; p50(m): Michal Zduniak/Shutterstock; p52(t): Mimafoto/Shutterstock; p52(m): Markus Mainka/Shutterstock; p52(b): Pixel-Shot/Shutterstock; p52(bl): Panther Media GmbH/Alamy Stock Photo; p52(br): Courtesy of the author; p54(t): John Birdsall/Alamy Stock Photo; p54(b): PHIL LENOIR/Shutterstock; p56(tl): Albert Russ/Shutterstock; p56(tm): Rvkamalov gmail.com/Shutterstock; p56(tr): Miro Novak/Shutterstock; p56(ml): xpixel/Shutterstock; p56(mr): Martyn F. Chillmaid/Science Photo Library; p56(b): GIPhotostock/Science Photo Library; p57: GIPhotostock/Science Photo Library; p58: Joe Gough/Shutterstock; p59: Jake Lyell/Alamy Stock Photo; p60: photopia/Shutterstock; p61(l): DKN0049/Shutterstock; p61(r): Sebastian Janicki/Shutterstock; p62(l): Aminadab Aldama/Shutterstock; p62(r): Andrzej Grygiel/EPA/Shutterstock; p63(t): DEA/S. VANNINI/Contributor/Getty Images; p63(m): Skoda/Shutterstock; p63(b): Evgeny Murtola/Shutterstock; p63(bl): Soumen Tarafder/Shutterstock; p63(bm): Trong Nguyen/Shutterstock; p63(br): ekipaj/Shutterstock; p64(l): manfredxy/Shutterstock; p64(r): Vladimir Tronin/Shutterstock; p66(t): MP_P/Shutterstock; p66(b): Middle East/Alamy Stock Photo; p67(l): Dovzhykov Andriy/Shutterstock; p67(r): Bukhta Yurii/Shutterstock; p68(t): Sam Wordley/Shutterstock; p68(bl): sukra13/Shutterstock; p68(bm): photocritical/Shutterstock; p68(br): Ody_Stocker/Shutterstock; p69(t): wasanajai/Shutterstock; p69(m): hxdbzxy/Shutterstock; p69(b): botazsolti/Shutterstock; p70(t): Sailorr/Shutterstock; p70(bl): Alexandru Rosu/Shutterstock; p70(br): Alexey Rezvykh/Shutterstock; p71: LuYago/Shutterstock; p72(t): ZAO2006/Shutterstock; p72(m): Pabkov/Shutterstock; p72(b): Rudy Umans/Shutterstock; p73(l): wassiliy-architect/Shutterstock; p73(r): Angrybirds/Shutterstock; p74: Voronin76/Shutterstock; p76(l): imnoom/Shutterstock; p76(m): Maria Uspenskaya/Shutterstock; p76(r): Albert Russ/Shutterstock; p77(tl): Science Photo Library; p77(tm): Lawrence Migdale/Science Photo Library; p77(tr): RHJPhtotoandillustration/Shutterstock; p77(m): Standard Studio/Shutterstock; p77(b): Science Photo Library; p78(t): Vladi333/Shutterstock; p78(b): Paceman/Shutterstock; p79: Andrew Lambert Photography/Science Photo Library; p80: GIPhotostock/Science Photo Library; p82(tl): GIPhotostock/Science Photo Library; p82(tr1): Standard Studio/Shutterstock; p82(tr2): Ody_Stocker/Shutterstock; p82(tr3): Standard Studio/Shutterstock; p82(tr4): Standard Studio/Shutterstock; p82(b): Ihor Matsiievskyi/Shutterstock; p83: Andrew Lambert Photography/Science Photo Library; p84(t): Purple Clouds/Shutterstock; p84(bl): Natthawon Chaosakun/Shutterstock; p84(br): Niko_V/Shutterstock; p85(t): keith morris/Alamy Stock Photo; p85(bl): nimon/Shutterstock; p85(br): Chad McDermott/Shutterstock; p87: FooTToo/Shutterstock ;p88(tl): Seika Chujo/Shutterstock; p88(tm): Fotofermer/Shutterstock; p88(tr): Michael Kraus/Shutterstock; p88(m): Idea tank/Shutterstock; p88(bm): Jerry Mason/Science Photo Library; p88(br): Andrew Lambert Photography/Science Photo Library; p88(bl): IanRedding/Shutterstock; p89: t_korop/Shutterstock; p90(l): mewaji/Shutterstock; p90(r): Jack Jelly/Shutterstock;

p91(t): AlexVector/Shutterstock; p91(ml): Bjoern Wylezich/Shutterstock; p91(mr): rukawajung/Shutterstock; p91(bl): Alexey Rezvykh/Shutterstock; p91(br): bilwissedition Ltd. & Co. KG/Alamy Stock Photo; p92: Luis Echeverri Urrea/Shutterstock; p93(t): QueSeraSera/Shutterstock; p93(b): Oleg Zaslavsky/Shutterstock; p94(t): Mastaco/Shutterstock; p94(m1): Daniel hughes/Shutterstock; p94(m2): Wanannc/Shutterstock; p94(m3): PTZ Pictures/Shutterstock; p96(t): amnsingh/Shutterstock; p96(ml): JUMBORUSHI/Shutterstock; p96(mr): atalavera/Shutterstock; p96(b): GIPhotostock/Science Photo Library; p97(t): Mary Terriberry/Shutterstock; p97(b): UniversalImagesGroup/Contributor/Getty Images; p98(t): ManuelSchafer/Shutterstock; p98(b): Andrew Lambert Photography/Science Photo Library; p99(t): Martyn F. Chillmaid/Science Photo Library; p99(b): Martyn F. Chillmaid/Science Photo Library; p102(l): Neil Fraser/Alamy Stock Photo; p102(r): cloki/Shutterstock; p103(t): Science Photo Library; p103(b): Charles O'Rear/Getty Images; p104(t): Catmando/Shutterstock; p104(b): Joaquin Corbalan P/Shutterstock; p105(t): Kolonko/Shutterstock; p105(b): hydebrink/Shutterstock; p106(t): arda savasciogullari/Shutterstock; p106(b): ODI/Alamy Stock Photo; p107(t): vchal/Shutterstock; p107(b): namu_zip/Shutterstock; p108(t): fboudrias/Shutterstock; p109(l): US Geological Survey; p109(r): e.backlund/Shutterstock; p111(t): Kingppin/Shutterstock; p111(b): R R/Shutterstock; p113: Ammit Jack/Shutterstock; p114: Stas Ponomarencko/Shutterstock; p115: ton koene/Alamy Stock Photo; p117(t): chemistrygod/Shutterstock; p117(b): magnetix/Shutterstock; p118(t): magnetix/Shutterstock; p118(b): Alon Za/Shutterstock; p119(t): Natata/Shutterstock; p119(b): ROMANVS Roman Mojzis/Shutterstock; p120(t): Science History Images/Alamy Stock Photo; p121(t): Fabrice Coffrini/Staff/Getty Images; p121(m): Photo by Michael Hoch; p121(b): SaraGiordano/Shutterstock; p124: Sebastian Janicki/Shutterstock; p125: MarcelClemens/Shutterstock; p126(ml): Morozov Anatoly/Shutterstock; p126(mr): Keith Homan/Shutterstock; p126(bl): Lyubov Timofeyeva/Shutterstock; p126(br): Laboko/Shutterstock; p128(l): Fedor Selivanov/Shutterstock; p128(r): Ava Kabouchy/Shutterstock; p129(t): shao weiwei/Shutterstock; p129(b): Maximchuk/Shutterstock; p130: Jordan Wende/Shutterstock; p131(t): P&F Photography/Alamy Stock Photo; p131(m): Helen Sessions/Alamy Stock Photo; p131(b): Joko P/Shutterstock; p132(t): Martyn F. Chillmaid/Science Photo Library; p132(b): Turtle Rock Scientific/Science Source/Science Photo Library; p133(l): Bjoern Wylezich/Shutterstock; p133(r): YuryKara/Shutterstock; p134(l): Joao Virissimo/Shutterstock; p134(r): elen_studio/Shutterstock; p136: Sean Sprague/Alamy Stock Photo; p137: RealityImages/Shutterstock; p138(t): maradon 333/Shutterstock; p138(b): Science Photo Library; p140(t): Madlen/Shutterstock; p140(b): sulit.photos/Shutterstock; p142: Rabbitmindphoto/Shutterstock; p145: Nasky/Shutterstock; p146: Lena Pan/Shutterstock; p147(t): 135pixels/Shutterstock; p147(b): Antonio Guillem/Shutterstock; p148(t): Charles D. Winters/Science Photo Library; p148(b): GIPhotostock/Science Photo Library; p149(t): Lawrence Migdale/Science Photo Library; p149(b): ruzanna/Shutterstock; p150: Photoongraphy/Shutterstock; p151(tl): golf bress/Shutterstock; p151(tr): Werayuth Tes/Shutterstock; p151(b): DimaBerlin/Shutterstock; p154(t): Maks Narodenko/Shutterstock; p154(b): Isaieva Liudmyla/Shutterstock; p156(t): roibu/Shutterstock; p156(m): Andrew Lambert Photography/Science Photo Library; p156(bl): Andrew Lambert Photography/Science Photo Library; p156(br): Andraž Cerar/Shutterstock; p157(l): Charles D. Winters/Science Photo Library; p157(m): GIPhotostock/Science Photo Library; p157(r): Chepko Danil Vitalevich/Shutterstock; p158(t): Andrew Lambert Photography/Science Photo Library; p158(b): Martyn F. Chillmaid/Science Photo Library; p159: Ivanov Andrey M/Shutterstock; p160(t): Axl4Real/Shutterstock; p160(b): Martyn F. Chillmaid/Science Photo Library; p161: Jose Arcos Aguilar/Shutterstock; p162(tl): Sergey Kamshylin/Shutterstock; p162(tr): Nneirda/Shutterstock; p162(bl): MikroKon/Shutterstock; p162(br): Perla Berant Wilder/Shutterstock; p163: ZHMURCHAK/Shutterstock; p164(tl): Juan Miguel Aparicio/Shutterstock; p164(tr): Andrew E Gardner/Shutterstock; p164(m1): Ody_Stocker/Shutterstock; p164(m2): Viktorija Reuta/Shutterstock; p164(b): Ody_Stocker/Shutterstock; p170: SpeedKingz/Shutterstock; p172(tl): RHJPhtotoandilustration/Shutterstock; p172(tr): Andraž Cerar/Shutterstock; p172(b): Albert Russ/Shutterstock; p174: Julia Kuznetsova/Shutterstock; p176(t): Peter Hermes Furian/Shutterstock; p176(bl): RHJPhtotoandilustration/Shutterstock; p176(bm): Ihor Matsiievskyi/Shutterstock; p176(br): Turtle Rock Scientific/Science Photo Library; p178(t): Victor1153/Shutterstock; p178(b): StudioMolekuul/Shutterstock; p179(l): DKN0049/Shutterstock; p179(tr): StudioMolekuul/Shutterstock; p179(b): DKN0049/Shutterstock; p180: feedbackstudio/Shutterstock; p181: Sebastian Janicki/Shutterstock; p182(l): S-F/Shutterstock; p182(r): Rob Crandall/Shutterstock; p184(t): Walter Eric Sy/Shutterstock; p184(b): olllikeballoon/Shutterstock; p185: Christa Fischer Walker; p187: StudioMolekuul/Shutterstock; p188(t): oatawa/Shutterstock; p188(m1): PRILL/Shutterstock; p188(m2): Nordroden/Shutterstock; p188(b): ThiagoSantos/Shutterstock; p190: Pat_Hastings/Shutterstock; p191(t): Olesya Kuznetsova/Shutterstock; p191(b): CKP1001/Shutterstock; p192(t): Dima Sobko/Shutterstock; p192(mr): Artisticco/Shutterstock; p192(ml): J. Palys/Shutterstock; p192(b): DmitrySt/Shutterstock; p194(tl): Joel_420/Shutterstock; p194(tr): asharkyu/Shutterstock; p194(m): Salar de Uyuni/Shutterstock; p194(b): Charles D. Winters/Science Photo Library; p196: geogif/Shutterstock; p197(l): Turtle Rock Scientific/Science Photo Library; p197(r): Martyn F. Chillmaid/Science Photo Library; p198: NASA/Chris Gunn; p202: PA Images/Alamy Stock Photo; p204(t): Science Photo Library; p204(b): Lawrence Migdale/Science Photo Library; p205: Charles D. Winters/Science Photo Library; p206: Kapuska/Shutterstock; p207: Andrew Lambert Photography/Science Photo Library; p208(t): Nick Poon/Shutterstock; p208(b): Henri Stierlin/Bildarchiv Steffens/Bridgeman Images; p209(t): Funtap/Shutterstock; p209(b): goran_safarek/Shutterstock; p210(t): Denis Zhitnik/Shutterstock; p210(bl): abriendomundo/Shutterstock; p210(br): ssuaphotos/Shutterstock; p211(t): Huguette Roe/Shutterstock; p211(m): MIKE MANIATIS/Shutterstock; p211(b): Vorotylin Roman/Shutterstock; p212(tl): Dimijana/Shutterstock; p212(tm): Martyn F. Chillmaid/Science Photo Library; p212(tr): Fablok/Shutterstock; p212(m): Viktorija Reuta/Shutterstock; p212(b): Viktorija Reuta/Shutterstock; p214(t): NGCHIYUI/Shutterstock; p214(b): Ihor Matsiievskyi/Shutterstock; p216: studiomirage/Shutterstock; p218(tl): Kenneth Sponsler/Shutterstock; p218(tr): Petrova Maria/Shutterstock; p220: Martyn F. Chillmaid/Science Photo Library; p222(t): Ari N/Shutterstock; p222(b): Martyn F. Chillmaid/Science Photo Library; p224: Crown Copyright/Health & Safety Laboratory/Science Photo Library; p225(l): Nataly Studio/Shutterstock; p225(r): hlphoto/Shutterstock; p228: Benedikt Juerges/Shutterstock; p229: tinkivinki/Shutterstock; p230(t): KGPA Ltd/Alamy Stock Photo; p230(bl): Breck P. Kent/Shutterstock; p231(r): Warpaint/Shutterstock; p231(l): United States Geological Survey; p232(t): B. Murton/Southampton Oceanography Centre/Science Photo Library; p232(b): NGDC/NOAA/Phil Degginger/Alamy Stock Photo; p234(t): neftali/Shutterstock; p234(b): Breck P. Ken/Shutterstock; p239(br): Catmando/Shutterstock.

Artwork by Integra Software Services, Q2A Media Services Pvt. Ltd, Erwin Haya, Wearset Ltd, Peter Bull Studios, Peter Stayte, Stéphan Theron, IFA Design (Plymouth, UK), and Clive Goodyer.

This Student Book refers to the Cambridge Lower Secondary Science (0893) Syllabus published by Cambridge Assessment International Education.

This work has been developed independently from and is not endorsed by or otherwise connected with Cambridge Assessment International Education.

Author's Acknowledgements

Enormous thanks to Barney, Catherine and Sarah Gardom for their sparkling suggestions, superb support and marvellous model making. Huge thanks to my parents, Mary and Edward Hulme, for patiently correcting my holiday diaries all those years ago, and getting me into writing from an early age. Big thanks to all at OUP, and to my editors. Finally, thank you to Shaun for selling me the best sit-stand desk ever, right in the middle of lockdown.

Every effort has been made to contact copyright holders of material reproduced in this book. Any omissions will be rectified in subsequent printings if notice is given to the publisher.

The periodic table of the elements

1 H Hydrogen																	2 He Helium
3 Li Lithium	4 Be Beryllium											5 B Boron	6 C Carbon	7 N Nitrogen	8 O Oxygen	9 F Fluorine	10 Ne Neon
11 Na Sodium	12 Mg Magnesium											13 Al Aluminum	14 Si Silicon	15 P Phosphorus	16 S Sulfur	17 Cl Chlorine	18 Ar Argon
19 K Potassium	20 Ca Calcium	21 Sc Scandium	22 Ti Titanium	23 V Vanadium	24 Cr Chromium	25 Mn Manganese	26 Fe Iron	27 Co Cobalt	28 Ni Nickel	29 Cu Copper	30 Zn Zinc	31 Ga Gallium	32 Ge Germanium	33 As Arsenic	34 Se Selenium	35 Br Bromine	36 Kr Krypton
37 Rb Rubidium	38 Sr Strontium	39 Y Yttrium	40 Zr Zirconium	41 Nb Niobium	42 Mo Molybdenum	43 Tc Technetium	44 Ru Ruthenium	45 Rh Rhodium	46 Pd Palladium	47 Ag Silver	48 Cd Cadmium	49 In Indium	50 Sn Tin	51 Sb Antimony	52 Te Tellurium	53 I Iodine	54 Xe Xenon
55 Cs Cesium	56 Ba Barium	71 Lu Lutetium	72 Hf Hafnium	73 Ta Tantalum	74 W Tungsten	75 Re Rhenium	76 Os Osmium	77 Ir Iridium	78 Pt Platinum	79 Au Gold	80 Hg Mercury	81 Tl Thallium	82 Pb Lead	83 Bi Bismuth	84 Po Polonium	85 At Astatine	86 Rn Radon
87 Fr Francium	88 Ra Radium	103 Lr Lawrencium	104 Rf Rutherfordium	105 Db Dubnium	106 Sg Seaborgium	107 Bh Bohrium	108 Hs Hassium	109 Mt Meitnerium	110 Ds Darmstadtium	111 Rg Roentgenium	112 Cn Copernicium	113 Nh Nihonium	114 Fl Flerovium	115 Mc Moscovium	116 Lv Livermorium	117 Ts Tennessine	118 Og Oganesson
		57-71 lanthanoids															
		89-103 actinoids															

57 La Lanthanum	58 Ce Cerium	59 Pr Praseodymium	60 Nd Neodymium	61 Pm Promethium	62 Sm Samarium	63 Eu Europium	64 Gd Gadolinium	65 Tb Terbium	66 Dy Dysprosium	67 Ho Holmium	68 Er Erbium	69 Tm Thulium	70 Yb Ytterbium	71 Lu Lutetium
89 Ac Actinium	90 Th Thorium	91 Pa Protactinium	92 U Uranium	93 Np Neptunium	94 Pu Plutonium	95 Am Americium	96 Cm Curium	97 Bk Berkelium	98 Cf Californium	99 Es Einsteinium	100 Fm Fermium	101 Md Mendelevium	102 No Nobelium	103 Lr Lawrencium

Note: the numbers shown are the proton numbers of the elements.